Springer Theses

Recognizing Outstanding Ph.D. Research

For further volumes:
http://www.springer.com/series/8790

Aims and Scope

The series "Springer Theses" brings together a selection of the very best Ph.D. theses from around the world and across the physical sciences. Nominated and endorsed by two recognized specialists, each published volume has been selected for its scientific excellence and the high impact of its contents for the pertinent field of research. For greater accessibility to non-specialists, the published versions include an extended introduction, as well as a foreword by the student's supervisor explaining the special relevance of the work for the field. As a whole, the series will provide a valuable resource both for newcomers to the research fields described, and for other scientists seeking detailed background information on special questions. Finally, it provides an accredited documentation of the valuable contributions made by today's younger generation of scientists.

Theses are accepted into the series by invited nomination only and must fulfill all of the following criteria

- They must be written in good English.
- The topic should fall within the confines of Chemistry, Physics, Earth Sciences, Engineering and related interdisciplinary fields such as Materials, Nanoscience, Chemical Engineering, Complex Systems and Biophysics.
- The work reported in the thesis must represent a significant scientific advance.
- If the thesis includes previously published material, permission to reproduce this must be gained from the respective copyright holder.
- They must have been examined and passed during the 12 months prior to nomination.
- Each thesis should include a foreword by the supervisor outlining the significance of its content.
- The theses should have a clearly defined structure including an introduction accessible to scientists not expert in that particular field.

Ghazi Ben Ayed

Architecting User-Centric Privacy-as-a-Set-of-Services

Digital Identity-Related Privacy Framework

Doctoral Thesis accepted by
University of Lausanne, Switzerland

Springer

Author
Dr. Ghazi Ben Ayed
Faculty of Business and Economics (HEC)
Department of Information Systems
University of Lausanne
Lausanne
Switzerland

Supervisor
Prof. Solange Ghernaouti
Faculty of Business and Economics (HEC)
Department of Information Systems
University of Lausanne
Lausanne
Switzerland

ISSN 2190-5053 ISSN 2190-5061 (electronic)
ISBN 978-3-319-38374-3 ISBN 978-3-319-08231-8 (eBook)
DOI 10.1007/978-3-319-08231-8
Springer Cham Heidelberg New York Dordrecht London

Printed on acid-free paper

Springer is part of Springer Science+Business Media (www.springer.com)

Parts of this thesis have been published in the following articles:

- Ben Ayed, G., Ghernaouti-Hélie, S., **Service-Oriented Digital Identity-related Privacy Interoperability: Implementation Framework of Privacy-as-a-Set-of-Services (PaaSS)** 2012 4th International IFIP Working Conference on Enterprise Interoperability (IWEI), pp. 193–200 http://link.springer.com/chapter/10.1007%2F978-3-642-33068-1_18
- Ben Ayed, G., Ghernaouti-Hélie, S., **Architecting Interoperable Privacy within User-Centric Federated Digital Identity Systems: Overview of a Service-Oriented Implementation Framework** 2012 4th International Conference on Networked Digital Technologies (NDT), pp. 165–177 http://link.springer.com/chapter/10.1007%2F978-3-642-30567-2_14
- Ben Ayed, G., Ghernaouti-Hélie, S., **Disassembling Digital Identity-Related Privacy into a Set of Services: SoaML-based Services Design** 2012 3rd International Conference on Exploring Services Sciences (IESS), pp. 44–57 http://link.springer.com/chapter/10.1007%2F978-3-642-28227-0_4
- Ben Ayed, G., Ghernaouti-Hélie, S., **Privacy Requirements Specification for Digital Identity Management Systems Implementation: Towards a digital society of privacy** 2011 6th IEEE International Conference for Internet Technology and Secured Transactions (ICITST), pp. 602–607
- Ben Ayed, G., Ghernaouti-Hélie, S., **Digital Identity Management within Networked Information Systems: From Vertical Silos View into Horizontal User-Supremacy Processes Management** 2011 14th IEEE International Conference on Network-Based Information Systems (NBiS), pp. 98–103
- Ben Ayed, G., Ghernaouti-Hélie, S., **Digital Identity Attributes Cohesion to Access E-services: Major Issues and Challenges in Digital Society** (2011) Journal of E-Technology, Volume 2, Number 3, pp. 89–97
- Ben Ayed, G., Ghernaouti-Hélie, S., **XRD Digital Identity Metadata-Based Approach to Foster Collaborations across Networked Computing Ecosystems** 2011 3rd International Conference on Networked Digital Technologies (NDT), pp. 105–119 http://link.springer.com/chapter/10.1007%2F978-3-642-22185-9_10
- Ben Ayed, G., **Digital Identity Metadata Scheme: A technical approach to reduce digital identity risks** 2011 International Workshop on Information Security and Risk management of the 25th IEEE International Conference on Advanced Information Networking and Applications (AINA), pp. 607–612
- Ben Ayed, G., **Consolidating Fragmented Identity: Attributes Aggregation to Secure Information Systems** 2008 IADIS International Conference on Information Systems, elected through blind review process as **Best Position Paper**

To the Lord Almighty

Supervisor's Foreword

As a professor in the Faculty of Business and Economics at the University of Lausanne, Director of the Swiss Cybersecurity Advisory and Research Group, I had the privilege and pleasure of supervising the doctoral research of Ghazi Ben Ayed, a body of research that led to a Ph.D. thesis entitled: "Architecting User-Centric Privacy-as-a-Set-of-Services: Digital Identity Related Privacy Framework."

In this work, Dr. Ben Ayed has developed a global and systematic approach to the problems of the management of digital identities and of maintaining confidence in the systems used for such identity management. He proposed elements of solutions for allowing digital identities and the parameters associated with these identities, essentially private data, to be managed by their owners and only made visible and available according to criteria defined by those owners. In doing so he contributed to the protection of personal data while considering the relevant technical and legal constraints.

Of particular note were his interdisciplinary approach, validated through the creation and verification of models for confidence and assurance, and his innovative approach towards proposing a technical solution to guarantee the right for data to be forgotten.

Along with the other members of the examining panel, I was struck by the quality of the research and of the practical solutions that were presented: these demonstrated a clear mastery of the conceptual principles of the field as well as of technical and technological matters. Specifically, Dr. Ben Ayed's research required finding answers to the central question of how to design and implement interoperable privacy systems based on the use of digital identities. The response consisted of developing a framework based on the needs for privacy created by the use of digital identities, designing Privacy as a Set-of-Services (PaaSS), and demonstrating how this could be implemented within the framework of Service-Oriented Architecture (SOA).

As a visionary in this field, Ghazi has been able to anticipate and pre-empt the description of the needs for the protection of personal data and of privacy in the age of the information society. With political, economic, legal, and technical stakes at play, the control of digital data has become a widespread desire. Real wars are breaking out around this new search for power and profits. Espionage, surveillance and the manipulation of information are current affairs and nobody can now be unfamiliar with the dangers linked to weaknesses in data protection. Through his work Ghazi ben Ayed demonstrates the existence of new possibilities for the owners of digital data to protect those data. He provides them with the means of re-establishing control over their own information assets and shows that it is not necessary to remain powerless in the face of the abusive and inappropriate use of our private data.

Lausanne, March 2014 Prof. Solange Ghernaouti

Preface

This work has been elected the best thesis in information systems in the information systems department, Faculty of Business and Economics, University of Lausanne, Switzerland (2012). It has also been nominated for European Research Consortium for Informatics and Mathematics (ERCIM) Best Ph.D. Thesis Award on Security and Trust Management (2013) and for Faculty's Outstanding Dissertation Award, Faculty of Business and Economics, University of Lausanne, Switzerland (2012). Additionally, the first published article of this work has been awarded "Best Position Paper" in one of the international conferences in information systems (2008).

We present the approach and results of work that has been conducted at the Department of Information Systems (ISI), University of Lausanne, Switzerland. We consider this work as a crutial step towards the realization of our **service-oriented cyber-security vision**: Could cyber-security be delivered a set of autonomous hosted services available per request on per-usage basis? We leave an increasingly digital footprint in cyberspace and this situation puts our digital identity at high risks. Privacy is a right and fundamental social value that could secure digital identities. Thus, the main question of this research is how to turn digital identity-related privacy in a shape of set of services that are loosely coupled, publicly hosted and available to on-demand calls. It is recognized that technical initiatives are not enough to guarantee resolution for the concerns surrounding a multifaceted and complex issue of identity and privacy. For this reason they should be apprehended within a global perspective through an integrated and a multidisciplinary approach, which dictates that privacy law, policies, regulations and technologies are to be crafted together from the beginning of the project as a set of requirements. They are drawn from global, domestic, and business-specific privacy laws and policies related to digital identity. We suggest a layered implementation DigIdeRP framework in accordance to model-driven architecture approach that would help cyber-security team to implement security requirements in the form of a set of services that could accommodate Service-Oriented Architecture (SOA): Privacy-as-a-Set-of-Services (PaaSS) system. The framework will serve as a basis for vital understanding between business management and technical managers on digital identity-related privacy initiatives. The layered framework presents five practical layers as an ordered sequence as a basis of security project roadmap, however, in practice, there is an iterative process to assure that each layer supports effectively and enforces requirements of the adjacent ones. Each layer is composed of a set of blocks, which determine a roadmap that security team could follow to successfully implement PaaSS. Several blocks' descriptions are based on OMG SoaML modeling language and BPMN processes description. We identified, designed, and implemented services that form PaaSS and described their consumption. PaaSS Java (JEE project), WSDL, and XSD codes are given and explained.

April 2014 Dr. Ghazi Ben Ayed

Acknowledgments

I would like to express my gratitude to every person who contributed and helped both directly and indirectly to make this doctoral project in the best conditions. In particular:

I would like to express my sincere gratitude to my Ph.D. supervisor Prof. Solange Ghernaouti-Hélie for the valuable guidance and advices. Her generosity and willingness to motivation contributed tremendously to the achievement.

I have been fortunate in having unwavering support of my mother Zahra and my father Abdelwaheb. My vocabulary fails me in thanking them for their guidance, understanding, and patience from my young age. I do warrant a special recognition.

I am grateful to the love of my life Nourchène, son Mohamed Reyam, and daughter Kenza for all the sacrifices. Heartfelt and endless thanks go to all my family members, in particular my beautiful sisters, brothers-in-law, and parents-in-law for their encouragement and support to pursue my interests. Words will never be strong enough to express my recognition and my deepest gratitude.

I want to do justice to my own teachers in helping me to broaden my view and knowledge. A special thank you to the teaching body of école primmaire El-Menzah5 and lycée secondaire El-Menzah6, Tunis, Tunisia; Institut Supérieur de Gestion-University of Tunis, Tunisia; McGill University, HEC Montréal, University of Montreal, Canada; and University of Lausanne, Switzerland.

Last, but not least, my gratitude is extended to all my research colleagues who continue to share their wealth of knowledge in order to make the digital world a safer environment and to turn it from an "un-forgetting" into a "forgiving" place.

Contents

Acronyms

BPMN	Business Process Model and Notation
CCNx	Content-Centric Networks
DigIdM	Digital Identity Management
DigIdDoc	Digital Identity Document
DigIdMeta	Digital Identity MetaEngine
DigIdeRP	Digital Identity-Related Privacy
FIM	Federated Identity Management
IdP	Identity Provider
IS	Information Systems
PaaSS	Privacy-as-a-Set-of-Services
PET	Privacy-Enhanced Technologies
SOA	Service-Oriented Architectures
SoaML	Service-oriented architecture Modeling Language
SP	Service Provider

Chapter 1
Introduction and Motivations

There is a powerful tension in our relationship to technology.
We are excited by egalitarianism and anonymity,
but we constantly fight for our identity.
David Owens (Professor at Vanderbilt University)

1.1 Context and Research Motivations

The advent of Internet-compliant technologies and open standards are easing the extension of information systems by lowering the barriers to connecting disparate business applications both within and across corporate boundaries. Increasingly, information technology architects are asked to define end-to-end business processes that span borders to enable inter-enterprise collaborations and mass integration with partners. Therefore, the current fortress landscape becomes a puzzle of partnering enterprises that should be working hand-in-hand toward building a common defense program in order to fortify the security of critical resources available within and across information systems [1]. Identity management systems span technological, political and social boundaries, and have become a strategic requirement for today's enterprise. Organizations could achieve both tactical benefits for the present and strategic benefits for the future. They can immediately benefit from regulations' compliance, such as privacy, security will be improved, fraud will be minimized and operating costs will be reduced [2]. Particularly, identity federation scheme, such as the Identrus consortium,[1] supports re-use of credentials and infrastructure to minimize cost and it supports the separation of authentication and attributes stores, allowing privacy and data control issues to be managed [3]. Thus, efficient management of digital identities is a critical need of the agile and profitable enterprise [4].

---

[1] http://www.identrust.com

G. Ben Ayed, *Architecting User-Centric Privacy-as-a-Set-of-Services*, Springer Theses,
DOI: 10.1007/978-3-319-08231-8_1, © Springer International Publishing Switzerland 2014

Identity and privacy are complex concepts and should be studied from different perspectives, thus, a multidisciplinary approach becomes a necessity. The complexity of managing identity and privacy comes from multiple reasons such as the nature of identity and privacy that have multiple facets: technological, social, legal, and cultural; and the fragility of digital identity bounded with immaturity of privacy in the digital life [5, 6]. Moreover, it is questioned whether information privacy and security are positively correlated in some situations and negatively correlated in others? And how stable or dynamic is the relationship between them is different technological settings and organizational environments? In addition and depending on the situation, users face identity retention and disclosure tradeoff. Sometimes, they are obliged to disclose digital identities but sometime users refrain from sharing digital identity to prevent possible exposure and privacy breaches; and in another side they disclose digital identity attributes and other information to make online transactions, seek convenience, and have fun. In the offline world, anonymous transactions can be conducted successfully, but in the service-oriented online world trust should be established between parties [7–9]. There are times when individuals need a secure and an accurate representation of themselves and other times when people may want to have the ability and freedom to project a quite different persona in online world to that in the offline world [10]. Moreover, these conflicting needs and requirements are compounded by a technological capability that is moving far too fast for society and companies to adapt to [11]. Additionally, the diversity of regulations and privacy policies rise trans-borders issues because they are set with different intents, purpose, and outcomes increases complexities [12]. Thus, a technical approach is not sufficient enough to tackle privacy issues and Privacy-enhanced Technologies (PET) is an example of technical initiative failure [10]. They have proved useful only in very narrow domains and did not respond adequately to the online world needs [13, 14]. A multidisciplinary and integrated approach dictates that law, policies, regulations and technologies are to be crafted together.

Internet is being criminalized. The fraudulent use of individual identity has increased at an alarming rate, thus privacy and identity management can play a key role to secure participation in digital society. Digital identity is bringing a whole new dimension to our existing identities. We leave an increasingly digital footprint in cyberspace such as digital records of our prenatal scans available on Flickr, personal profile within a social networks, death information in FamilySearch[2] historical records, data collected by diverse agencies on our behalf, blogs' contributions, emails, performed searches with various engines. Trails are memorized by the network, while, in most cases, we still don't have the capabilities to delete them if we wish. Major online service providers memorize, access, and exploit 'Web of trails' for their own commercial benefits, and as a result, we are losing control over our personal data and leaving our identity at a high risk. One hundred million worldwide Facebook users are threatened by identity theft as

[2] http://fsbeta.familysearch.org/

a repercussion of Facebook hack case [15], in which personal details have been collated and published on file-sharing service. The dramatic increase in identity theft and other types of digital identity is unlikely to end soon. Security, identity theft, incorrect computer records, credit rating destruction, privacy, online purchasing and banking, loss of identity, misuse of personal information, phishing, identity cards, behavioral monitoring and tracking, etc. The list of concerns is long and people still feel concerned and worried about the digital world, security and loss of control. Criminal forces have organized themselves internationally to trick users into releasing valuable information through phishing schemes, to inadvertently install spyware in users' computers and harvests information through pharming attacks, or to stealing a vast amount of identities by targeting corporate, government and educational databases. Criminal networks are working toward acquiring and reselling identities and the international character of these networks makes them increasingly difficult to penetrate and dismantle. Privacy is a critical right and protection to enforce, if we wish to provide to individuals with the means to secure and control their digital identities, while enabling organizations to exploit fairly this invaluable source of information. When privacy is compromised, security of the individual, the organization or the country could be threatened [7, 10, 11, 16–21].

Identity and privacy should be interoperable and distributed through the adoption of service-orientation and implementation based on open standards. Identity functionality is increasingly delivered as sets of services, rather than monolithic applications. It is hard to create an identity layer for the internet mainly due to the little agreement on what it should be done and how it should be run. The lack of agreement arises because digital identity is contextual in nature. Thus the emergence of a single simplistic universal digital identity solution is not realistic [17]. Privacy is to be engineered to integrate identity from the start, rather than attaching it to identity after the fact. It is confirmed that building secure systems requires privacy principles/policies to be taken into consideration from the early stage [2, 22]. Design must start from maximum of privacy is one of the design principles of European PRIME Project [23]. Organizations are realizing that they need better security, particularly identity and privacy management through a better interoperability both within and between countries. Interoperability is not just technical interoperability but the alignment of policy, services and processes with business requirements [24]. W3C Platform for Privacy Preferences (P3P) Project is a step towards interoperability by making privacy policies of web sites transparent for automated agents but the use of SSL to protect connections to public sites and deployment of Kerberos within enterprises lacked global vision and they've been implemented only for specific domains. Service-Oriented Architectures (SOA) is widely used in distributed and dynamic systems and driving a loosely coupled approach to application interoperability and integration [25]. We borrow OMG SoaML SOA definition: "SOA is a way of describing and understanding organizations, communities and systems to maximize agility, scale and interoperability". SOA defines how people, organizations and systems provide and use services to achieve results [26]. In system that is implemented following the SOA approach, functionalities are delivered and consumed as services [27]. SOA aims to simplify development and delivery of new business functionalities, enabling reusability and interoperability. Thus, services would be built according to a prescribed set of standards, protocols, and interfaces, which make them interoperable and reusable [28]. Thus, an identity layer in which identity and privacy management services are loosely coupled, publicly hosted and available to on-demand calls could be more realistic and an acceptable situation.

Digital identity management projects requires a set of guidelines and advices [18], Oracle suggested best practices and SOA governance framework [29] to help make SOA implementation projects. Thus, there is a need to build a framework to better manage implementation risks and encourage stakeholders work together, collaboratively throughout the process as a team. The framework allows people, processes, and technology to be collaboratively integrated [30].

1.2 Problem Statement and Research Outcomes

In this thesis, we aim to respond to the following main questions: how identity architects and designers could design interoperable digital identity-related privacy system? Other questions are also important to respond in order to be able to answer the main research question: how to capture business interoperability described in the form of digital identity-related privacy (DigIdeRP) requirements? How to disassemble business interoperability into set of services (technical interoperability): Privacy-as-a-Set-of-Services (PaaSS) system? The research is information system design-type in the field of security and its outcome is to suggest a layered framework to help security implementation team to design, architect, and implement PaaSS system. The framework relays on the idea that privacy requirements should be taken into consideration from the beginning of system development project and privacy regulations/policies could be incorporated into technology. Service-oriented architecture modeling language (SoaML) diagrams are used to convert requirements into set of services.

Digital identity attributes are supposed to be shared after setting up a contract between parties. In Fig. 1.1, the subject asks for a service from the service provider (SP), which gives back a digital identity-related privacy contract form. The subject chooses to accept terms of privacy contract then the SP asks the required digital identity attributes. The subject replies by specifying Identity Provider(s) IdP(s) that SP should reach. Involved IdP(s) contact the subject in order to receive digital identity attributes release confirmation. The subject confirms, IdP(s) release attributes and SP gives the service to the subject. Various SPs and IdPs are distributed within a circle-of-trust (discontinued line eclipse) and collaborate with the subject to deliver the service. Parties involved in circle-of-trust should have been already agreed to comply with terms of privacy contract. Such distributed environment imposes the need of interoperability to execute and apply terms and

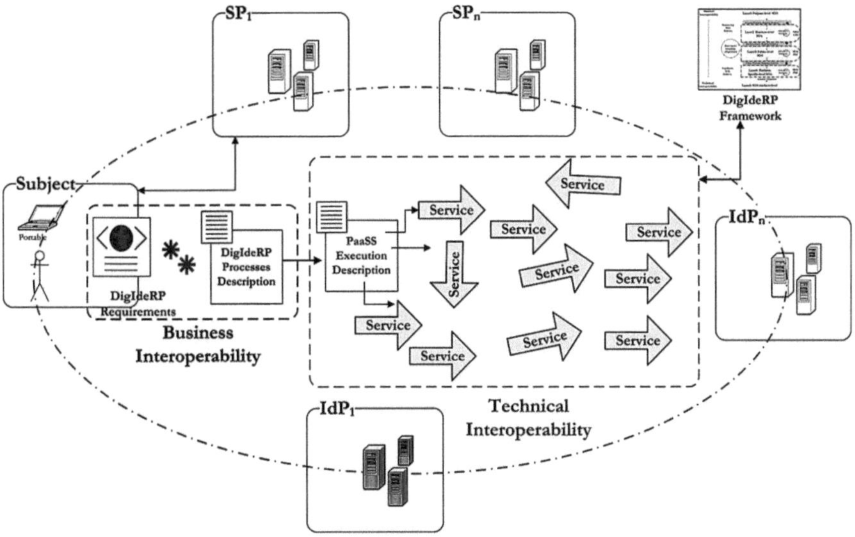

Fig. 1.1 Technical and business interoperability

conditions of privacy contract between parties. With the emergence of service-oriented architecture and open standards as means of interoperability, we suggest a five-layer DigIdeRP implementation framework to disassemble terms and conditions of privacy contract into a set of collaborated services: Privacy-as-a-Set-of-Services (PaaSS) system.

Following steps of the framework, we began with identification of business interoperability, through the definition of DigIdeRP requirements that are drawn from global, domestic and business-specific privacy policies. DigIdeRP requirements enumerate a set of objectives capable of being widely enough accepted to serve as backplane for distributed systems. Because these requirements are drawn from major privacy policies, they reflect a remarkable convergence of interests and organizational will to implement them. Each requirement ends up giving rise to an architectural principle guiding the construction of PaaSS. The framework is not only a technical-view framework, rather, it is multidisciplinary and multiple views framework that gather different roles and responsibilities in implementation security team. Top level security management is responsible for specification of the purpose-level SOA (layer1); security/privacy business analysts are responsible of business-level SOA (layer2); security/privacy architects are responsible of fabric-level SOA (layer3); and security/privacy systems developers are responsible of platform-specific-level SOA (layers4) and SOA-Artifacts-level (layer5). Mapping gateways ensure the transition between two layers, thus layers' owners have to collaborate and communicate to successfully conduct the mapping. Mapping gateways help to avoid siloed implementation and assure a shared effort. Flow chart diagrams and documents could facilitate the communication between owners and contribute to the success of the mapping.

DigIdeRP Framework helps to align DigIdeRP initiatives with organization's business goals and security strategy. Such initiative requires an engagement from top level security management throughout the project. The framework's components are distributed over five layers and three mapping gateways to define the roadmap that security implementation team should follow to successfully conduct the project. The framework allows not only service identification, design, and implementation but also service executions to support DigIdeRP requirements translated into BPMN business processes. The framework is enough flexible to allow multi-perspectives services implementation. It allows implementing services based on range of perspectives: network operator centric perspective, application service provider centric perspective, or user-centric perspective. In each perspective, we should describe the requirements in the form of conversation and information exchange between SP, IdP, and Subject. Even if the DigIdM technical model is not identity federation, centralization could be a good candidate, see Chap. 3. Because it is built in accordance to model-driven approach, the framework should accelerate the implementation because it could be supported by a range of design and implementation tools in order to have automatic code generation.

1.3 Thesis Outline

After introducing the thesis by setting the scene, describing research motivations and justifications, and specifying the research question, we provide high-level dissertation structure and a brief summary of the major contributions in each chapter as follows. We discuss in Chap. 2 multiple facets and fundamentals of digital identity and describe major issues and complexities surrounding digital identity. However, in Chap. 3, we provide taxonomy of digital identity management (DigIdM) definitions based on three types of definition-focus: technical, management, and user-supremacy. We explain that DigIdM should have a horizontal process view and service orientation. We provide a description and comparison between DigIdM technical models and we give supremacy to digital identity federation and particularly to its derivate user-centricity. We propose an innovative approach based on Metadata usage to make less visible persistent digital identity documents, thus, users would be given more control over digital identity information. We implement this approach on Content-Centric Networks (CCNx). In Chap. 4, we discuss the basics of privacy and issues surrounding digital identity-related privacy. We study and group privacy policies into three policy classes: global, domestic and business-specific privacy policies. We draw DigIdeRP requirements from these privacy policies related to digital identity. Ten DigIdeRP requirements are identified: purpose specification of attributes collection, consent for attributes usage and release, limited usage of attributes, limited retention of attributes, accuracy of stored attributes, openness, authentication and enrollment needs, choice and terms of the contract, secondary use, and compliance. These requirements will be considered as a starting point to implement target's Privacy-as-a-Set-of-Services system. We provide, in Chap. 3, an overview of the Service-oriented Architecture (SOA) foundations and we explain DigIdeRP Framework in accordance of model driven engineering approach to implement PaaSS system, a technical interoperability. Such implementation requires business interoperability: DigIdeRP requirements. The requirements are described on business processes basis with Business Process Model and Notation (BPMN). Six DigIdeRP processes are identified and explained. We choose OMG Service Oriented Architecture Modeling Language (SoaML) to identify and describe the pool of autonomous, granular and loosely coupled services. The BPMN processes description combined with SoaML services' description allows defining service consumption roadmap. We present in Chap. 6 SoaML design toolkit and SOA artifacts of the user-centric digital identity federation participants (SP, IdP, and Subject). Few corresponding pieces of codes are given with explanations. Finally, a brief summary conclusions and main research contribution are included in Chap. 7. We identify research limits and several areas of future research work and improvements.

References

1. G. Ben Ayed, Consolidating fragmented identity: attributes aggregation to secure information systems. IADIS Int. J. Comput. Sci. Info. Syst. **4**, 1–12 (2009)
2. P. Mackinnonm, in *Large-Scale Identity Management in Digital Identity Management: Perspectives on the Technological, Business and Social Implications*, ed. by D.G.W. Birch (Gower Publishing Limited, Farnham, 2007), pp. 105–112
3. J. Skipper, in *Authentication in Business in Digital Identity Management: Perspectives on the Technological, Business and Social Implications*, ed. by D.G.W. Birch (Gower Publishing Limited, Farnham, 2007), pp. 95–104
4. A. Scorer, in *Identity Directories and Databases, in Digital Identity Management: Perspectives on the Technological, Business and Social Implications*, ed. by D.G.W. Birch (Gower Publishing Limited, Farnham, 2007)
5. H. Noonan, B. Curtis, "*Identity*", in *Stanford Encyclopedia of Philosophy* (2009), Available: http://plato.stanford.edu/entries/identity/
6. E. Dallaway, Loss of privacy: internet security's high price. Infosecurity Magazine **4**(7), 10 (Elsevier, 2007)
7. P.J. Windley, *Digital Identity: Unmasking identity management architecture* (IMA) (O'Reilly Media, California, 2005)
8. M. Benantar, *Access Control Systems: Security, Identity Management and Trust Models* (Springer, Berlin, 2006)
9. S. Clauβ, M. Köhntopp, Identity management and its support of multilateral security. Comput. Netw. Int. J. Comp. Telecommun. Netw. **37**(2), 205–219 (2001)
10. International Telecommunication Union, Digital life. ITU internet report. (2006), Available http://www.itu.int/osg/spu/publications/digitalife/docs/digital-life-web.pdf. Accessed 21 May 2010
11. P. Cochrane, in "*Forward of the Book,*" *in Digital Identity Management: Perspectives on the Technological, Business and Social Implications*, ed. by D.G.W. Birch (Gower Publishing Limited, Farnham, 2007)

12. G. Hosein, Politics of information society: the bordering and restraining of global data flows. (2004), Available: http://www.privacyinternational.org/survey/censorship/unesco.pdf. Accessed 6 May 2010
13. D.G.W. Birch, N.A. McEvoy, in "*A Model for Digital Identity," in Digital Identity Management: Perspectives on the Technological, Business and Social Implications*, ed. by D.G.W. Birch (Gower Publishing Limited, Farnham, 2007), pp. 95–104
14. F. Stalder, The failure of privacy enhancing technologies (PETs) and the voiding of privacy. J. Sociol. Res. **7**(2) (2002)
15. Facebook 'hack' releases 100 million user details onto filesharing sites. Inforsecurity USA. (2010), Available: http://www.infosecurity-us.com/view/11343/facebook-hack-releases-100-million-user-details-onto-filesharing-sites/. Accessed 30 Nov 2010
16. S. Philippsohn, in *ID and the Law, in Digital Identity Management: Perspectives on the Technological, Business and Social Implications*, ed. by D.G.W. Birch (Gower Publishing Limited, Farnham, 2007), pp. 193–203
17. K. Cameron (ed.), *The Laws of Identity* (Microsoft Corporation, Washington, 2005)
18. M. Hansen et al., Privacy and identity management. IEEE Secur. Priv. **6**, 38–45 (2008)
19. G. Bell, J. Gemmel, A digital life. Sci. Am. Mag. **296**, 58–65 (2007)
20. K. Cukier, *A Special Report on Managing Information* (The Economist, London, 2010). (23 Feb–5 Mar)
21. Organizing Committee of Digital Identity and Privacy (Human Capital and Social Innovation Technology Summit), Call for contribution to managing digital identities for education, employment and business development. (2007), Available: http://events.eife-l.org/HCSIT2007/overview/dip/dip2007. Accessed 11 May 2010
22. Center for Democracy and Technology, Privacy principles for identity in the digital age [draft for comment—version 1.4]. (2007), Available: http://www.cdt.org/files/pdfs/20071201_IDPri vacyPrinciples.pdf. Accessed 28 May 2010
23. PRIME Community. PRIME—privacy and identity management for Europe document, (2005), Available: https://www.prime-project.eu/ Accessed 14 Feb 2010
24. J. Elliott, in *Planning ID Management in Government, in Digital Identity Management: Perspectives on the Technological, Business and Social Implications*, ed. by D.G.W. Birch (Gower Publishing Limited, Farnham, 2007), pp. 183–191
25. P. Harding, An introduction to identity-enabled web services. (2005), Available: http://wp.bitpipe.com/resource/org_1138132830_877/9180_IndentityWebServ_edp.pdf?site_cd=bp. Accessed 15 Nov 2011
26. OMG, Service oriented architecture modeling Language (SoaML)—specification for the UML profile and metamodel for services (UPMS). (2009), Available: http://www.omg.org/spec/SoaML/1.0/Beta2/PDF/. Accessed 20 June 2010
27. T. Varvarigou, V. Andronikou, in *IDIS '09 Identity Management in GRID Computing and Service Oriented Architectures: Research and practice* Presented at the Second Multidisciplinary Workshop on Identity in the Information Society, (London, 2010)
28. G. Cernosek, An introduction to architecture management. (2006), Available: http://www.ibm.com/developerworks/rational/library/dec06/cernosek/index.html. Accessed 15 Nov 2011
29. M. Afshar et al., SOA governance: framework and best practices. (2007), Available: http://www.oracle.com/us/technologies/soa/oracle-soa-governance-best-practice-066427.pdf. Accessed 25 March 2011
30. D. Kelley, Practical approaches for securing web applications across the software delivery lifecycle. (2009), Available: https://www14.software.ibm.com/iwm/web/cc/imc/rational/papers/security-pafswa/SecuringWebApplicationsintheSDLC.pdf. Accessed 17 Sept 2010

Part I
Cyber-security

Chapter 2
Digital Identity

An identity is questioned only when it is menaced.
James Baldwin (1924–1987), American novelist
and civil rights activist

2.1 Introduction

Having an identity and expressing it have been of that importance from the early time. Inscribed ostrich shell fragments found in Diepkloof Rock Shelter in Western Cape, South Africa are among the earliest examples of the use of symbolism as a form of expressing identity.

Figure 2.1 shows three over 270 pieces of decorated shells are dated to about 60,000 years ago, 20,000 older than cave painting, which was considered presently the first form of writing in history. At that time the ostrich eggs were used as bottles once engulfed their content. The researchers, who have investigated the material since 1999, argue that the markings are almost certainly a form of messaging—of graphic communication [1–4]. Dr. Pierre-Jean Texier from University of Bordeaux, France explains: "the lines are crossed at right angles or oblique angles by hatching. By the repetition of this motif, early humans were trying to communicate something. Perhaps they were trying to express the identity of the individual or the group" [1]. In the ancient near east excavation brought to light a group of clay tablets and wooden boards, dating to the middle of the third millennium B.C., on which Sumerian and Akkadian inscribed identity information of the collectivity such as geographical names, names of gods, names of rulers, names of exorcist, and hymns, legal documents, medical records, and lists of professions. In addition, they wrote in colophons individuals' names such as authors and tablets' collectors [5, 6].

G. Ben Ayed, *Architecting User-Centric Privacy-as-a-Set-of-Services*, Springer Theses,
DOI: 10.1007/978-3-319-08231-8_2,

Fig. 2.1 An old representation of the identity of an individual or a group [1]

2.2 Identity: Yesterday and Today

The term 'identity', which is firstly known used in 1570, has been used in many different ways in academic research and in popular usage [7]. The term is still of disputed origins, but it's certainly true that the its origin derives from Middle French 'identité', from Late Latin 'identitat-, identitas', or probably from Latin 'identidem' repeatedly, a contraction of 'idem et idem' and literally 'same and same' [8]. In the American Heritage Dictionary of the English Language, the term 'identity' could refer to 'the collective aspect of the set of characteristics by which a thing is definitively recognizable or known', 'the set of behavioral or personal characteristics by which an individual is recognizable as a member of a group', 'the quality or condition of being the same as something else', 'the distinct personality of an individual regarded as a persisting entity', or 'information, such as an identification number, used to establish or prove a person's individuality, as in providing access to a credit account' [9].

In the pre-modern times, human identity was defined by geography, community, and family relationships. If an individual was born into a well-known and rich family in London, that is typically the environment in which he or she would remain. If an individual began life in a poor remote community in India, they would typically not be able to change their life pattern or economic status over time. One's geophysical space and one's place in society were inextricably linked, the possibility of freedom of movement being severely limited. With modern times there arrived a greater choice for participation in different social circles, and the possibility of social and economic mobility. Today, most people carry some form of identification on them at all times, but this practice is relatively recent in human history. In the past, the declaration of an individual's name, sometimes accompanied by the name of their city or village, was sufficient to prove their identity. This is no longer the case. Further, the notion of identity today can refer not only to humans, but extends to animals, machines, and other objects or resources. A machine may have an identity which would allow it to access certain information at certain times, or be employed by some individuals, to the exclusion of specified others [10].

2.3 Identity Perspectives: Multiple Facets of the Identity

For many centuries, stories of the holy fool Mulla Nasrudin's have been studied in Sufi circles for their hidden wisdom [11]. One of the stories tells of Mulla Nasrudin who traveled to another city. Before he left on his journey, his wife put a sign around his neck with his name on it so that he would not forget his identity. In his way, he spent a night at a caravanserai; while he slept, a joker took the sign and put it around his neck. When the Mulla awoke, he was appalled to find his name tag on the joker's chest. He cried: "It seems that you are me. But if you are me, then who am I?" [12, 13]. The Mulla's dilemma is a ridiculous one, but it illustrates the importance of identity and introduces the multiple perspectives aspect in studying the identity. The Mulla question is one of the key questions in the philosophical debate over identity. The Mulla dilemma touches one of the central identity-related issues in social, cultural, and child and adolescent sciences. In addition, the story could illustrate the importance of losses related to identity theft in digital economy and e-commerce.

Identity concept is seen from different perspectives and applicable into different domains. We describe here multiple perspectives of the identity and mention few major issues from each perspective. The identity debate dates back to ancient world's philosophy. In general, personal identity in philosophy is employed referring to Who am I? It consists roughly of those properties that make the individual unique and different from others [14]. From the same perspective, identity refers to a set of qualities and characteristics that make an entity definable, distinguishable, and recognizable comparing to other entities. In recent times, many philosophers have given attention to the question of change impact over time on the personal identity continuity such as Aristotle that distinguished between 'accidental' and 'essential' identity changes. Accidental changes refer to identity properties changes such as hair color change, while essential changes are radical and don't preserve the identity like someone, who dies. Other concepts arise such as numerical and qualitative identities. Numerical identical is the same one: one thing rather than two, but qualitative identical is exactly similar two things such as twins. The 'personal identity' has addressed the conditions to stay numerical identical throughout time [10, 15]. 'Identity formation' is defined as the process of the fabrication of the distinct personality of an individual in a particular stage of life such as establishment of a reputation. In this context, pieces of identity include a sense of personal continuity, a sense of uniqueness from others, and a sense of affiliation. These pieces could help people to define their selves in the eyes of others and themselves [16]. From the mathematical perspective, the law of identity in logic is upheld by a reflexive relation, states that an object is always the same as itself ($A = A$). In mathematics, the term identity denotes several meaning. Specifically, in algebra, the identity function $ids(x) = x$ for all x in the set S. The identity matrix includes ones on the main diagonal and zeros elsewhere. In social science, we use the term identity referring to an individual's comprehension of himself as a discrete, separate entity [17]. From the legal perspective, protection policies of sensitive identity-related information policies are critical, and privacy regulations are on the rise. Although from a technology viewpoint, the priorities may be authorization and control, what seems to be different and evolving is the notion of equipping the end user with the necessary controls to protect his identity information: Users are informed about what data is requested from them and how their personal data is treated, e.g. for what purpose it is used and who can access. Through this process, users can decide whether to provide their data and to consent to the service provider's data handling policies. Ideally, the service provider employs technical components such as access control systems to enforce the consented policies; for instance, to ensure that a user's e-mail address is not used for marketing but only for the consented billing purpose. From the cultural perspective, cultural identity deals with the influence of an individual identity by his belonging to a group or a culture [18]. Other questions and issues arise such as ethnicity, citizenship, nationhood, and how culture could influence on emotion, thoughts and self. In his book 'Culture and Identity' [13], Charles Lindholm states that since the late nineteenth century, psychological anthropology scholars study of the relationship between the individual's identity and culture. The discipline addresses also fundamental questions about the nature of humanity that have become pressing in the present era of multiculturalism and globalization. In social sciences, identity is a modern formulation of dignity, pride, and honor. One of the key question related to identity in social sciences is 'Who is we?' referring to the concept of social identity complexity [19]. It deals with an individual's subjective representation of the interrelationships among his multiple social group identities. The same authors mention that membership in many different groups, multiple social identities, can lead to greater social identity complexity, which can foster the development of global identity. From the economic perspective, particularly in marketing, a corporate identity is visibly manifested by the use of trademarks and the way of branding. Corporate identity is established when there is a common ownership of corporate philosophy, values, and norms that help the attainment of business objectives [20]. In their book titled 'Identity Economics' [7], the authors demonstrates how identities shape the employees' work, wage, and well-being. In psychology, a 'psychological identity' is related to self-image, self-esteem and individuation. It might be defined as a network of values and convictions that structure the individual's life. Moreover, it considered also as a property or a set of properties that an individual might have for a while and then lose, thus, he would acquire a new identity or perhaps carry on without one [10]. The family therapist and child psychiatrist Salvador Minuchin provides psychological definition of identity. He declares "the human experience of identity has two elements: a sense of belonging and a sense of being separate" [21]. From computer science and information technology perspectives, digital identity, online identity and others concepts have emerged. We cover these aspects further on this dissertation. From history, anthropology, and archeology perspectives, identity refer to human origins and identity construction over time. From genetics perspective, major issues are addressed such as genetics and origins of species, and how molecular genetics influence human personalities. From the art perspective, we mention architecture and identity such as architecture in Islamic culture. We mention the religious perspective to point people may see their identity as defined partly by some moral or spiritual commitment such as Islamic, Catholic, Jewish or anarchist. Or they may define it in part by the nation or tradition they belong to as an Armenian or a Québécois [22]. Finally, from the political perspective, many ongoing debates are over ethnic, race, gender [23], national, and transnational identities [24]. From the sociological perspective, the author [25] provides definitions and the distinctions of 'identity' and 'identification' concepts. 'Identity' denotes the ways in which individuals and collectivities are distinguished in their relations with other individuals and collectivities; and 'identification' is the systematic establishment and signification, between individuals, between collectivities, and between individuals and collectivities, of relationships of similarity and difference.

2.4 Digital Identity: Definitions, Basics and Nomenclature

Digital identity is composed of two distinct words that we explain each one separately: (1) 'identity' is what makes individuals the same today as they were yesterday (sameness), but it is also what makes them different from one another (uniqueness). Though these fundamental concepts have remained the same over time, changes in economic and social structures have affected the determination and perception of identity. Identity is the distinction between the private and the public spheres of human existence, and as such identity and privacy are forcibly linked [10]. As the boundary between the private and the public in the digital age becomes increasingly blurred, the creation and maintenance of secure identities online has emerged as an important priority for businesses and consumers alike. The researchers [26] define 'identity' as a set of personal information and identity management system as authentication and attribute management system. While, [27] defines the identity establishment concept as 'the representation of methods by which, a user, a running process, or a thread of execution is securely associated with a legitimate entity'. The author states that the goal of 'identification and authentication (I&A)', which is the process of establishing a user identity, is to provide to the entity access only to authorized computer resources. However, [28] restrict the entity definition to people or organization and define the identity, within a specific application domain, as an entity representation through a generation of a unique key, which combines all the elements of identity information. The researchers [26] define 'identity' as a set of personal information and identity management system as authentication and attribute management system. While, [27] defines the identity establishment concept as 'the representation of methods by which, a user, a running process, or a thread of execution is securely associated with a legitimate entity'. The author states that the goal of 'identification and authentication (I&A)', which is the process of establishing a user identity, is to provide to the entity access only to authorized computer resources. However, [28] restrict the entity definition to people or organization and define the identity, within a specific application domain, as an entity representation through a generation of a unique key, which combines all the elements of identity information; and (2) in the Webster's New Explorer Dictionary, the word 'digital' means 'done with the finger or toe' and narrowly, a 'digital computer' is a mean by which 'provides a readout in numerical digits'. In today's ordinary technological parlance, 'digital medium' refers to machines that are capable of recording, transmitting, or receiving data in binary digit form. In addition, people are getting connected by consuming an increasing amount of digital media and broadband technologies, such as internet and mobile phone. We present in next sections a literature review of the definitions, basics, and preliminaries of digital identity. Digital life is designated to represent a daily life where individuals use digital mediums and technologies to engage activities in online and offline worlds. In the entitled Digital Life Internet Report [29] published by International Telecommunication Unit (ITU), the United Nations specialized agency for Telecommunication, experts in policy and strategy state that today's digital world is transforming individual lifestyles. Always-on internet access has become a global norm and daily lives has brimmed with SMS, e-mail, chats, multiplayer online gaming, virtual worlds and digital multimedia. But what does it mean digital, digital media, digital world, etc.?

Several definitions of the term 'digital identity', from different perspectives, have appeared in the literature. A simple definition is related to one of identity. Thus, identity is defined as a collection of data about subject that represent attributes, preferences, and traits [29], so in parallel, in the digital world a person's identity is typically referred to as their digital identity [29]. The term 'digital identity' has emerged through the evolution of the Internet. Wherever we go, we leave traces of fragmented information about our identity. Leaving a comment in a forum, filling out a form, maintaining a blog, creating a full profile (photo, name, phone number, etc.) in a social network, conducting a parallel existence, we are educating others about what we are, what we do and especially what we think and then constructing 'digital identity'. Internet users are striving to share their digital identity with others to re-enforce their online presence and one of the favorite users' activities on the net is egoGoogling. A 'personhood' means that we recognize that an entity or individual has a person's status and the 'digital personhood' means the person's status projected in digital environment [30]. The authors [31] suggest a conceptual definition of the term 'digital identity'. It refers to two concepts: 'nyms' (called also masks or aliases) and 'partial identities'. In his book [29], Windley defines a digital identity as the data that uniquely describes a subject or an entity and the ones about the subject's relationships to other entities'. The author gives the car title as example of digital identity. The car title contains vehicle identification number that uniquely identifies the car to which it belongs and other attributes such as year, model, color and power. The title contains also relationships such as the set of car owners from the time it was made. From technical perspective, the same author explains that digital identity is built on a set of technologies that includes cryptography, authentication, authorization, identity provisioning, directories, digital rights management, identity federation, and interoperability standards. In contrast, the author [27] does not distinguish between identity and digital identity. He provides a broad definition of identity from a computing perspective as 'a computer representation of an active entity that can be physical (such as human, a host system, or a network device) or a programming agent'. In the lexicon [32], the authors coincided digital identity and identity definitions as 'a representation of a set of claims made by one party about itself or another data subject' but the authors of Princeton University Wordnet [33] don't distinguish between the two concepts by arguing that either in the real or electronic worlds, an individual may have multiple identities. The same authors point out that identity entails 'individual characteristics by which a person is recognized or known' [33]. The authors of the definitions paper of OECD [30] report insist on the difference between the two concepts by defining the 'identity' as 'a limited notion of set of claims', whereas the 'digital identity' as 'a thing or an artifact that refers to a person'. Adam's speech and Adam's ID card are two claims of the same individual. Based on works of Jenkins [25] and Goffman [34], Professor Shirley Williams of the University of Reading, UK [35] distinguishes also between the identity as 'a social performance' and digital identity as 'performances in digital places', which means the persona that an individual presents across all the digital spaces. He explains that human identity is naturally social and always involves, in addition to agreement and disagreement, convention and innovation, communication and negotiation, a performance, which denotes the activity of an individual which occurs during a period. He highlights that digital reputation and trust are other people's interpretation of the person's digital identity [35]. Moreover, the authors [30] highlight the referential and partiality natures of identity. Referential because claims must refer to a person and partial identity refers to 'a subset of identity information as the thing may not be sufficient to identify a person at different moments in time'. They add that the term 'digital identities' is a synonym of 'partial identities' in which a set of identity attributes are enclosed [30]. Digital identity is considered as an intersection of identity and technology in the digital age [36]. The author of the Digital Identity book [29] points out that identity is crucial to enable the virtual 'place'. He adds that digital identity will ensure that internet infrastructure respond to multiple needs including security, privacy, and reliability.

The world of digital identity has its own nomenclature. The following terms are derived from [27, 28], Windley's book [29], SAML-OASIS glossary [37], Liberty Alliance Technical glossary [26, 38, 39]. An 'entity' represents an active element of a computer/network system. It could be a single person; a group of persons, an automated process, a set of processes, a software program, a subsystem, an entire organization, a machine, a host system, a networking device or in general other thing making a request to access a resource. An entity's access to a system is encapsulated an 'account' and the 'principal' is the internal representation of an active entity in a specific environment. 'Attributes' describes a property associated with the subject such as physical trait, network address, medical record, purchasing behavior, bank balance, credit rating, dress size, and age. Attributes can also include preferences and traits. 'Preferences' represent desires such as preferred seating on an airline, brand of ice cream, and preferred language, and used currency. 'Traits' are like attributes but two differences are noticed between them: traits are inherent rather than acquired, and attributes may change but traits change slowly. Examples of traits are person's blue eyes, hair color, company's location and date when it was incorporated. Since the distinction between attributes, preferences, and traits rarely makes a difference in the design of an identity system, we will typically use, in this dissertation, attribute to mean all three unless there's a need to distinguish among them. Attributes are often represented as pairs of attribute name and attribute value(s) and might be conveyed through an 'attribute assertion'. An 'Attributes Authority' (AA) manages the identity store and provides to IdP the requested attributes in the desired format such as through an attribute assertion. An 'identity store', 'repository' or 'directory' refer to any technology that could be used to store identity attributes such as the LDAP directories, databases, and files. Attribute 'scheme' or 'schema' represents the definition of the structure and the form of attribute held in a directory or database. 'Enrollment' is the process by which an identity of entity is created in a specific identity system. A 'Service Provider' (SP) interacts with entities primarily via HTTP and provides service to the user through a medium such as a portal (e.g. an online retailer, a financial institution, a government agency). An 'Identity Provider' (IdP) provides identity attributes to other providers (e.g. telecommunication company) and it may act as an authentication service provider. Note that 'provider' can refer to either SP or IdP and could interact and discuss details behind authentication. 'Attribute aggregation' is the ability to collect user attributes from IdP(s). An 'identifier' is used in two senses: (1) one that identifies; (2) uniquely refers to the system entity. Essentially, an identifier is a distinguished attribute of an entity. 'Credentials' are transferred data in order to establish a claimed entity identity and they allow transferring trust between subjects. 'Identification' is the process of using claimed or observed attributes of an individual to infer who the individual is. An 'identifier' points to a subject and it could be a name, a serial number, or some other pointer to the individual being identified. 'Pseudonym' is a name or label that may identify an individual within a system but does not correlate to that individual outside of the system. 'Secondary use' of information represents any use of identity or linked information that is inconsistent with an identity system's purpose. 'Authentication' is the process of establishing confidence in the truth of a number of claims. Finally, the following definitions drawn from the glossary of terms and definitions of the Ofcom research report [40]. 'Avatar' is defined as 'a computer user graphical representation of him or herself. An avatar can be two or three-dimensional'; a 'Profile' as 'the personal homepage on a social network site, usually including information about a user, photos, and their friend list. Profiles form the basis of social networking sites'.

2.5 Digital Identity, Security and Trust

Digital identity related mechanisms are the core of modern systems, networks, and applications security. In the book [27], the author considers that anonymity is not a desired computing goal but secure identification of users is the core element of computing security. He adds that the level of security is attached to an authenticated identity associated with it. The ultimate goal is to enable deterministic accountability and lay the foundation for responsible and secure computing [27]. Narrowly, identities are critical to define access control polices [41]. Identity is having more importance in the online world. In the offline world, anonymous transactions can be conducted successfully, but in the service-oriented online world, we have to know something about the service recipient. Building digital identity infrastructures is an attempt to establish a community of trust, which becomes a requirement for conducting online business [29]. For instance, eBay community of trust lays on users' reputations. Windley [29] points that in order to make use of digital identity; organizations are required to understand other concepts such as trust and privacy. Corporations are considering identity infrastructure to provide security so that interactions with customers, partners, employees, and suppliers become are more flexible and richer. The business should not be limited to just transactions, but relationships with customers, employees, suppliers, and partners and identity tends to change this relationship from one-way to a more customized one. Therefore, agile, business-responsive IT infrastructure should have at its core a flexible, interoperable identity infrastructure.

2.6 Digital Identity: Major Issues and Complexities

There is no a single problem of personal identity,
but rather a wide range loosely connected questions.
Stanford Encyclopedia of Philosophy

We do not intend to cover in this section all the issues related to digital identity rather than pointing couple of major issues and complexities.

2.6.1 Mutation from One YOU to Multiple YOUs

Currently, people are maintaining multiple identities. From the social science perspective, the recognition of an individual has no one, 'personal self', but rather 'several selves' that correspond to widening circles of group membership. Thus, an important issue that has been addressed is how individuals combine these different identities when they want to define a subjective identity within a social group? [10] Currently, the latter question becomes applicable to the digital/online world and being subject of many studies and researches. The authors [42] mention that the online world encapsulates a growing amount of scattered and unordered fragments of users' identities due to two major reasons. The first is because of the lack of a robust generic identification system and the second is the intentional creation of users' alternate identities. Figure 2.2 is an illustration. Creating more than one identity can be desirable for users depending on the context. A user may wish to be aggressive and egotistical in online multiplayer war game, but sensitive and sociable for virtual encounters and social networks. Thus, the online world represents an ideal nameless and faceless environment for users to easily create multiple representations of their identities:

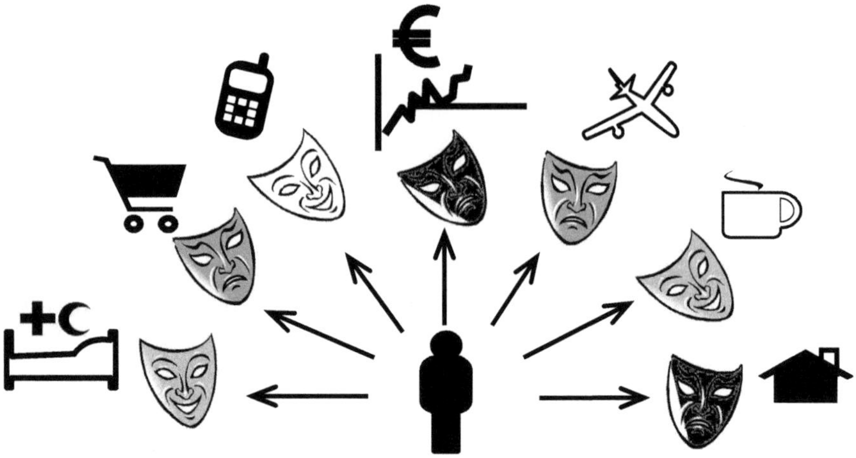

Fig. 2.2 Digital masks and partial identities

'digital personae' [42]. However, we usually speak of identity in the singular but in reality it is plural because it encapsulates multiple identities, 'perspectives', or 'facets' [29]. Researchers at Stanford University's Virtual Human Interaction Lab don't distinguish between 'digital you' and 'virtual you' and they consider them as synonyms of digital clone, avatar, nyms [43], personae [42], which strongly influences the 'real you' [44]. In contrast, the authors [45] defines the 'virtual you' as a representation of a virtual version of the subject in the virtual world.

In ITU 2006 "Digital Life" report [43], the authors mention that 'nyms' and 'profiles' provide the subjects interacting capabilities with other parties in different environments. For example, nyms enable subjects to exercise their freedom anonymously in digital life by setting up synthetic personae complete with attributes such as age, race or religion. Another example is 'social profiles' that are created in popular social Web sites and online networks such as MySpace,[1] Bebo,[2] and Facebook[3] could be useful by allowing the users to post and share content, and staying in touch with others. Actually, the users log-in with pseudonyms in order to preserve anonymity that what make these networks attractive but in the other side anonymous users could engage malicious activities.

Avatars could enable online interaction and business opportunities. An avatar is 'a graphical personification or incarnation of a user in a shared virtual reality space, more specifically, in online role-playing games and virtual universes (e.g. Second Life[4] and Active Worlds[5]) for a specific objective' [43].

---

[1] http://www.myspace.com

[2] http://www.bebo.com

[3] http://www.facebook.com

[4] http://www.secondlife.com

[5] http://www.activeworlds.com

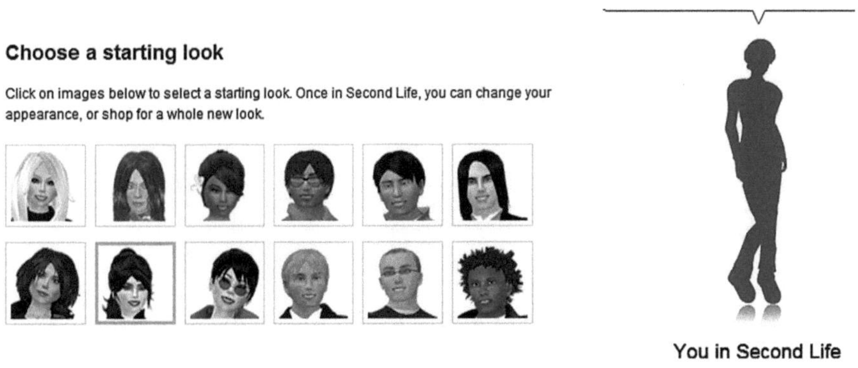

Fig. 2.3 Selecting the right 'you in second life' (avatar)

In the Second Life, abbreviated "SL", the user can choose multiple avatars with speech and language capabilities, Fig. 2.4, to participate within different virtual situations, such as virtual meeting, virtual tutoring and virtual commerce using virtual currency Linden Dollar (L$). The use of avatars has been extended to online social networks and forums and is affecting the identity construction such as the phenomenon of gender switching when the user uses opposite sex avatars [43]. In his book 'Coming of Age in Second Life' [46], the anthropologist Tom Boellstorff stresses the important role that avatar plays in everyday activities in SL. He says: "a man spends his days as a tiny chipmunk, elf, or voluptuous woman. Another one lives as a child and two other persons agree to be his virtual parents. Two "real"-life sisters living hundreds of miles apart meet every day to play games together or shop for new shoes for their avatars. The person making the shoes has quit his "real"-life job because he is making over five thousand U.S. dollars a month from the sale of virtual clothing" [46]. Besides providing a comprehensive introduction to social, economic, political, and cultural settings in which the new media operate, the author of the book [47] presents multiple reasons why people might take the opportunity to explore different identities, including: (1) the ability to change character and physical traits at will, as illustrated in Fig. 2.3. This will provide to users the opportunities to explore other forms of existence and change the ways in which they may be perceived by others; (2) the opportunity for shy people or those who are uncomfortable with face-to-face interaction to form relationships and express views freely; (3) the potential to bring geographically and socially disparate individuals together based on common interests, thereby stimulating dialogue and curbing loneliness. An avatar could represent the offline personality of the user or another more desirable personality that the user cannot construct and afford in the offline world. In the online world, contact with strangers is encouraged and expected. It is acceptable to exaggerate, hide, alter or undermine the truth about oneself in order to encourage constructing desirable online impressions or reputations. In the paper titled "the connected identity" [48], the author confirms the presence of a relationship between the image of the visual interface, such as visual pseudo or avatar, and the

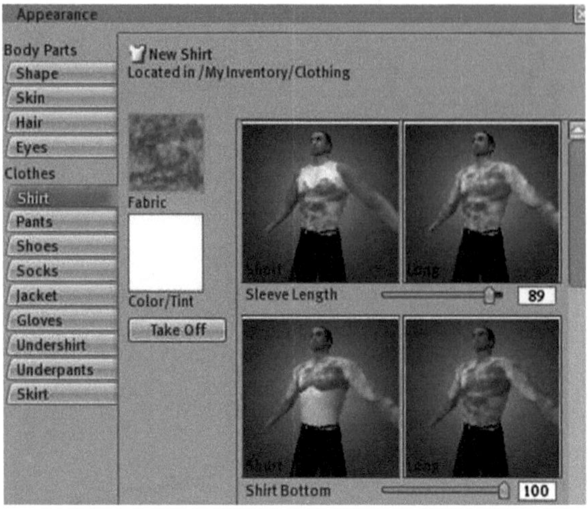

Fig. 2.4 Customizing the "appearance" of the second life's avatar

personality of the individual. He explains that the image reflects the personality of the user and shows who he really is. Identity has been building on freedom of expression through various media (videos, photos, blogs, avatars, music) and, so far, it is gaining upper hand over the sense of the community belonging [48].

Distributed attributes, which represent multiple set of different attributes within different environments, is a consequence of a context-based nature of identity concept. Partial identities [49] or digital selves [50] are any subset of attributes associated with entity that the entity itself can select for interacting with other parties. In the real-life, various forms of identity are required for various contexts in which the identity is to be presented in a suitable way and within suitable information by the identity holder [49]. For instance, the person A, as a traveler, is asked to provide a passport at the counter of customs or immigration to proof his identity; the person A, being a car driver, is asked to show his driving license to a police officer, who stopped him in the highway; the person A, as a customer, is asked to provide his credit card with fidelity saving card in a movie store to take advantage of DVD prices reductions; the person A, as a student, is requested to show his student card to have access to computer lab facilities; the person A, as a patient, is asked his medical card in the hospital to receive health services. Thus, for each domain, a specific partial identity is provided for identification. The partial identity can be named or unnamed, which means it might or might not be related to the entity's true identity. In order to establish trust between parties in the digital world, a subset of digital identity attributes needs to be communicated. Digital identities exist in specific contexts and the contextual relationship between them is crucial to managing transactions and interactions. The context will determine which subset of attributes is required, or which "partial identity" will establish enough trust for the transaction to go forward. At the organizational level, identity attributes are distributed over different environments such as files, enterprise directories, databases, and online social networks [49]. Similarly, in the online world, the authors [42] mention that users easily create multiple representations of their identities called 'digital personae'. The author [29] calls multiple identities or 'personas' that the subject holds as digital identity 'perspectives' or 'views', which represent different perspectives on who is the subject is and what attributes he processes. They represent also a set of attributes that other entities have and can access. For instance, a bank sees account attributes, a physician in a hospital sees health record attributes, a district police sees criminal records attributes, and the employer sees other attributes such as full name, social security number, and one bank account number for paycheck deposit. The same author, in his book [29], explains that identity attributes are classified and organized into three sets, called 'tiers of identity' [51]. Each identity tier maintains relationships with other and its perceived value by the subjects is different. Tier1, labeled 'My Identity', includes attributes and traits associated with the subject such as hair color. 'Shared Identity' is the label of tier2, which consists of the attributes that are assigned to the subject by others in the sake of identifying him temporarily within a specific context and based on some kind of relationship. Driver's license, credit card, health insurance card, library card are all examples of shared identity. Once the relationship that defines the identity is terminated (when the context changes), the attributes associated with it are no longer useful. Tier3 is largely about profiling; it deals with 'Abstracted Identity', which establishes abstractly the identity of a group. Marketing companies provides abstracted identity by classifying a subject as a male over 50, a Swiss Air frequent flyer, and a Geneva resident. The same author points that commonly the subject perceives the value and benefit of the tier2 identity relationship, which is usually established with his consent to meet a real need, however, tier3 relationships are usually forced on us and they rarely meet a real need. He states unsolicited commercial email or spam as a tier3 identity issue. The same author stresses that major identity issues that face organizations deal with tier2 identity. He stresses that how employees and customers perceive the effort to build digital identity infrastructure depends on their sensitivity to tier3 identity issues and their satisfaction towards the added-value of tier2 identity relationships. Tier2 relationships are dictated by organizations and consented by the individual. The one-way relationship is likely to change as service-oriented economy emerges. The power-shifts are brought on by increasing available services and improved systems that make it easier for customers to switch their allegiances; and more customized services would make it more likely that customers dictate their terms in their relationships. More specifically, the power-shift is the switch from the world of "take-it-or-leave-it" to "mass customization". Good business would recognize theses shifts. The author mentions two fallouts from the identity power-shift: (1) identity aggregation: multiple tier2 relationships create identity silos. From the user perspectives, multiple identities create inconvenience to maintain these identities but the user is generally willing to have his identities aggregated for more convenience in getting the desired services; (2) convergence of tier2 and tier3 identities: since the world is moving from mass marketing towards individual-specific marketing effort, there are chances that demographic groups related identity, tier3, will converge to tier2 identity [29]. Thus, identity fusion and aggregation takes a place.

Fig. 2.5 Identity views

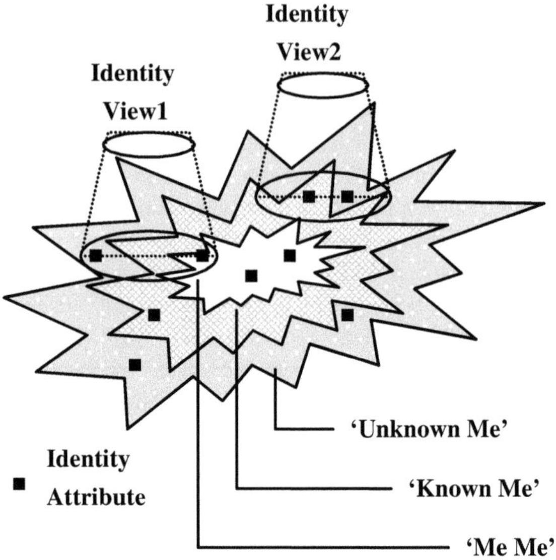

The authors [52] address multiplicity of 'views' or 'perceptions' that can exist on subject's identity in almost the same way to identity tiers. A single view defines a subject's digital identity that has a context's validity and appropriateness as shown in Fig. 2.5, which is adapted from [52]. There are three views of subject's identity: 'Me Me' refers to the part of the identity information that the person is aware of and directly controls (e.g. residence address). 'Known Me' is the part of identity information that the person is aware of and indirectly controls (e.g. revenue data and the associated tax levels that are under the control of the department). 'Unknown Me' is the part of identity information that the person is not aware of and over which the person has no control. This information can be controlled by known parties (e.g. certification authority) or by unknown parties (e.g. credit rating agencies and identity thieves) [52]. We believe that this picture of identity that comprises multiple views, perspectives, or views is derived from a multi-dimensional classification of the human world, and the definition and role of identity in social sciences. It is said that: "identity is to know 'who's who' (and hence 'what's what'). This involves knowing who we are, knowing who others are, them knowing who we are, us knowing who they think we are, and so on: a multi-dimensional classification of the human world and our places in it, as individuals and as members of collectivities" [53].

We believe that multiple YOUs constitute the identity, or overall identity, of the subject. We borrow the words of Amin Maalouf, who grew up in Lebanon and now lives in France. He is the author of the book: "In the Name of Identity: Violence and the Need to Belong". He shares his perspective and answers the question about identity; is he considering himself half French and half Lebanese? He says "not at all! The identity cannot be compartmentalized; it cannot be split in halves or thirds, nor have any clearly defined set of boundaries. I do not have several identities; I only have one, made of all the elements that have shaped its unique proportions" [24].

2.6.2 Origins of Fragmented Identity

Digital identity is bringing a whole new dimension to our existing identities. We leave increasingly digital footprints in cyberspace forming a web of trails. Examples are digital records of our prenatal scans available on Flickr,[6] personal profile within a social networks, death information in FamilySearch[7] historical records, data collected by diverse agencies on our behalf, blogs' contributions, emails, performed searches with various engines. Visible or invisible, left consciously or not, the data aggregation contributes to the definition of our identity. Editing our personal profile within social networks is different from that carried out by an employer 'googling' of a prospective employee, tracking our activities as a citizen, and possibly inferring health problems from our undertaken activities in self-advocacy groups [50].

Friends or other people opinions about an individual are highly affecting his digital identity. For instance, social networks users can tag friends through free online tagging services such as TagMyPals.[8] Such service offers a set of predefined digital representations or avatars based on classification of people personalities. TagMyPals users can tag friends full names on the avatars based on their perception of others' personalities. The avatars and tags can be easily added to photos section in Facebook and Myspace. Above, in Fig. 2.6, few TagMyPals avatars. Distributed fragmented identity attributes is a consequence of a context-based nature of identity concept. In the real-life, various forms of identity are required to various contexts in which, the identity is to be presented in a suitable way and within suitable information by the identity holder. For instance, the person A, as a traveler, is asked to provide a passport at the counter of customs or immigration to proof his identity; the person A, being a car driver, is asked to show his driving license to a police officer, who stopped him in the highway; the person A, as a customer, is asked to provide his credit card with fidelity saving card in a movie store to take advantage of DVD prices reductions; the person A, as a student, is requested to show his student card to have access to computer lab facilities; the person A, as a patient, is asked his medical card in the hospital to receive health services [54]. The online world encapsulates a growing amount of scattered and unordered fragments of individuals' identities due to two major reasons [42]. The first is because of the lack of a robust generic identification system and the second

---

[6] http://www.flickr.com/

[7] http://fsbeta.familysearch.org/

[8] http://www.tagmypals.com

Fig. 2.6 Free tagging service according to friend's personality's classification

is the intentional creation of users' alternate identities. Creating more than one identity can be desirable for individuals depending on the context. A user may wish to be aggressive and egotistical in online multiplayer war game, but sensitive and sociable for virtual encounters and social networks [42].

Different enterprise directories store different pieces of identities. Modern organizations become distributed and maintain multiple identity repositories. This reality promotes spreading identity attributes across information systems and landscaping identity silos. Thus, different pieces attributes of our identity are contained in different environments such as files, enterprise directories, databases, and online social networks. We illustrate identity silos shaping and origins with the following use cases: (1) managing finance and preserving privacy. Rather than using a single credit card for shopping, most of the people prefer to use multiple credit cards to better manage finances and assure anonymity. A man buys a birthday's gift for his spouse with one of his credit cards rather than using the jointly held credit account. Therefore, each credit card issuer maintains a different set of user attributes; (2) managing attributes schema and policies restrictions. The restriction occurs when a number of identity stores do not allow write permission for several reasons, such as technical, governance and political reasons. In addition, the directory schema could be static and cannot be changed without major repercussions on the whole infrastructure. Hence, attributes would be stored only in a limited number of repositories and could not be distributed over all identity stores. We can extend this use case to point out that having identity attributes within different semantics, such as languages and cultural considerations could foster the identity fragmentation; (3) context-based nature of identity and governance issue. Each context requires a specific form of attributes to authenticate an identity holder; (4) technological advent and emergence. The identity management and access control related technologies have evolved within different computing waves that range from mainframes, mid-size systems to personal computing, and from enterprise distributed network infrastructure to the internet and web. The history of computing shows that new fragmented identities are created with the emergence of each discipline; (5) business dynamics. As a consequence of corporate mergers and acquisitions over time is a complex fragmented identity infrastructure; (6) Simple authentication and access management. Often, different lines of business or divisions maintain separate identity repositories in order to easily manage users' access to different and heterogonous business applications such as CRM and HR; (7) multiple Web subscription. Many web sites require user subscription before providing services. As a result, a growing array of online fragmented identities is maintained by the Web sites' back-ends [54]. Concurrently, in information systems, access control and policies are different within different applications. Each application or service provider requires a specific set of attributes to let the user access the assets. A person may hold multiple credit cards issued by multiple banks that results multiple set of client attributes distributed over multiple repositories and locations. Furthermore, each individual has a couple of static attributes such as date and place of birth and dynamic attributes that may change such as blood pressure, home address, and phone number. Thereby, each person would have multiple sets of different attributes within different environments [49, 54].

2.6.3 Digital Identity and Digital Memories

All of the person's communications with other people and machines, as well as the images he sees, the sounds he hears, the Web sites he visits, and the Web searches he performs are recorded. US president Barack Obama provided some counsel for youngsters who want to grow up and be president. He replied to a 9th grader at Wakefield High School in the Washington suburb of Arlington, Virginia, who asked how he too could become President one day, saying that: "When you're young, you know, you make mistakes and you do some stupid stuff (…) I want everybody here to be careful about what you post on Facebook, because in the YouTube age whatever you do, it will be pulled up again later somewhere in your life" [55]. In the Digital Life article of the Scientific American Magazine [56], Gordon Bell and Jim Gemmell state that human memory can be maddeningly elusive and the era of digital memories is inevitable. Recently, a team at Microsoft Research Labs has developed a system, called MyLifeBits, to mainly digitally chronicle every aspect of a person life and to provide some of the tools needed to compile a lifelong digital archive. When the person is on the go, the system continually uploads his location from a portable Global Positioning System device. All of these recording are transmitted and stored in a personal digital archive that is both searchable and secure. After 6 years, more than 300,000 records, taking up about 150 GB are amassed. Portable sensors can take readings of things that are not even perceived by humans, such as oxygen levels in the blood or the amount of carbon dioxide in the air. Sensors can also log the three billion or so heartbeats in a person's lifetime. The authors explain of the new systems services by saying: "New systems may allow people to record everything they see and hear—and even things they cannot sense—and to store all these data in a personal digital archive" [56]. The same authors questioned why recording someone's life becomes possible today than before. The author cites three main reasons: (1) the growth of digital storage capacity has been staggering. Today a terabyte (one trillion bytes) hard drive can store everything the person read including emails, Web pages, papers and books, all the music the person purchased and downloaded, 8 h of speech and

10 pictures a day for the next 60 years. The author predicts that if current trends continue, in 20 years, with the same hard drive price, a person can buy a 250 TB of storage. This capacity should be able to satisfy anyone's recording needs for more than 100 years; (2) some of these devices can record a wealth of information about the users; (3) the dramatic increase in computing power has led to the introduction of processors that can efficiently retrieve, analyze and visualize vast amounts of information. Metadata such as the date, place and subject of a photograph or written or spoken comments that the database appends to the file, are easing the retrieve, or recall, process of digital memories. However, the advent of the digital-memories era will not be trouble-free. Many countries currently impose restrictions on recording conversations or photographing people. Moreover, many individuals are equally concerned about recording information for three reasons: (1) information could be used against them in court; (2) information could invade privacy; and (3) fear of access to records by identity thieves, gossipmongers or authoritarian states. In addition, from the security perspective, storing a lifetime of personal data in a single archive is vulnerable. One of the major advantages of digital memories is also mentioned. Digital memories allow vividly reliving an event with sounds and images, enhancing personal reflection in almost the same way that the Internet has aided scientific investigations. Every word one has ever read, whether in an e-mail, a document or on a Web site, can be found again with just a few keystrokes [56]. Emmpanuel Hoog, the CEO of INA, answers the questions of Le Nouvel Observateur reporters about the future of the world's digital memory and how to civilize Internet. He explains that years ago, individual or collective memory, is considered as a rare cultural asset and therefore valuable. A 100 years ago, a family life was illustrated by a dozen of pictures. But today we take hundreds of photos in summer holidays with small digital camera and mobiles. We are passing to future generations a huge stock of digital memory. In addition, museums, archives, universities, heritage institutions have long been in charge of sorting and organizing knowledge, but today, nobody can accept this because each digital producer manage by himself his memory with his manner. This would weaken our ability to draw a common destiny. He adds that given the ever growing content available on the Internet, the fundamental issue is how to sort, to make choices. The government has focused so far on the issue of digitization of content but now it should focus on how to make content accessible to more people. He thinks also that authorities should urgently address the issue of access criteria and the hierarchy of knowledge on the web at local and regional levels. Today, the monopoly of access is between the hands of search engines, which are using non transparent criteria for web content indexing. Such content is considered as an economic asset, he urges public authorities to create real spaces for public service, knowledge and expertise on the Internet. Hood calls continuity logic between souvenirs, memory, and history as 'memorial ecosystem'. He adds that the memorial ecosystem is called into a question with the advent of the digital world. Yesterday, there was some continuity between stages and each stage is the pre-cursor of the next one. Today we can remember everything, thus souvenirs and memory are taking precedence over history. And somehow, too much

memory kills the memory-or, rather, too much memory kills the history. The explosion of the memorial bubble may produce two consequences: (1) the resurgence of the wars of memory. Because history can be unfair, every minority can claim its history and identity at a large scale. The excess of such claims may generate identity crisis; (2) amnesia and collective cultural loss; (3) why indeed memorize, since the machine remembers us? Hoog writes "always more memory, but still less marks" to explain that the right to forget he is needed as a requirement for democracy. Today, there is a tremendous privatization of our personal data in the Internet. Companies are drawing profiles on personal information of each of us. Despite the efforts of the National Commission for Informatics and Liberties (CNIL) in France, the situation is not satisfactory. Every citizen is in danger of his past that can reappear at any moment. At the same time, we become producers of memory and we have accepted a regression of our privacy. However, privacy, rights to privacy is the foundation of a liberal society. Hoog adds that the digital native would have the challenge how to search on the internet. In the real world everyone can distinguish with the naked eye a grocery store, a school, a town hall, and a garage. For the Internet, it should be the same thing. I think that civilized Internet is allowing everyone to navigate easily. It is a challenge that calls for new forms of public regulation. Not everything can be left to the search engines that are now the only players in the web, which structure and organize it [57].

2.6.4 Digital Identity in Social Networks

Social networking sites are gaining more and more importance on people daily life. They offer people ways to communicate and socialize with each other via the internet through a PC or mobile phone. Individual's friendship chain become part of digital identity. Would you be my online friend? Once the user finds a profile of a friend or someone else, he can add him by sending a message to the other user requesting friendship. If the recipient approves the connection, the relationship is visible through both users' list of friends. The friends' list typically includes a list of links to other friends' profiles. Thus, when participants surf on social network sites, they can jump from one profile to another through a friendship chain. Based on a research results published in the report [40], the average adult social networker has profiles on 1.6 sites, and most users check their profile at least every other day. Part of the digital identity is constructed through the web of trails that individuals are leaving in the online world, especially in social networks. In fact, thirteenth century Mulla's dilemma touches the central social problematic of identity construction [13] and, in the same way, the author of the book [7] explains that digital identity is bounded, not only to identity attributes, but to the individual's behavior. Thus, in a restricted manner, digital identity is bounded to individuals' behaviors in social networks. Trails could be customized profile information, opinion sharing about a subject or other friends, photos posting, and so on. The Ofcom report states also that users create well-developed profiles as the basis of their online presence and such profiles often contain very detailed individual's information, even though it is not compulsory to provide that much of information [58]. People could easily and simply create their own online page or profile, and construct and display an online network of contacts, often called 'friends' [40]. As examples of well-known social networks: Friendster,[9] MySpace,[10] Facebook,[11] Bebo,[12] Skyrock Blog,[13] Hi5,[14] Orkut,[15] LiveJournal,[16] and CyWorld.[17] In order to join these networks, a user should register and create a social profile by entering a set of static and dynamic user attributes such as their demographics and tastes, a self-description, and often photos that provide a visual image. The participant's social profile is considered in this context as a social persona being a part of his digital identity. Some social networks sites allow participants to articulate and publicly display their relations to others in the system, which, in turn, allow viewers to traverse the network.

Online social profiles and activities is having more visibility and gaining more accessibility through "Universal Social Networks", abbreviated "USN". USN, called also social networking convergence service [58], is basically an application which focuses on making easier for end users to create content independently of the blogging platform usage. It allows updating all the blogs and web services from within one environment. USN permits to make it easy for the end user to let his friends and colleagues around the web know what he's up to and what he's writing. It keeps the end user friends on any network informed about his activities. But what are the consequences of USN usage on our digital identity? We present a list of USNs that are classified into four categories: (1) Social feed aggregator, called also lifestreem, or online presence aggregators: MyMashable [59], Profilactic [60], Snag [61], Profileomat [62], Naymz [63], SocialURL [64], PeopleAggregator [65], ProfileFly [66], SocialNetwork.in [67], and Mashable [68]. These services are ready to exploit but others are still in status of work in progress such as ProfileLinker, Upscoop, MyLifeBrand, Tabber, Ex.plode.us, Correlate.us, Istalkr, and SocialStream [69]; (2) desktop aggregator, an application that provide a single access to many social networks and aggregation capabilities: 8hands [70, 71], NoseRub [72, 73], and Minggl [74, 75]; (3) people finder such as Wink [76], a people search over the user profiles of MySpace, LinkedIn and Bebo. Spokeo [77] is another example of people finder that offers a search, by name, email address, phone number and friends; and (4) users' bookmarks aggregator such as SecondBrain [78].

---

[9] http://www.friendster.com

[10] http://www.myspace.com

[11] http://www.facebook.com

[12] http://www.bebo.com

[13] http://www.skyrock.com/blog

[14] http://www.hi5.com

[15] http://www.orkut.com

[16] http://www.livejournal.com

[17] http://us.cyworld.com

2.6.5 *Digital Identity, Context-Awareness, and Ubiquity*

Establishing the identity of a person is becoming an important need in context-aware environments. Context awareness originated as a term from ubiquitous computing, called also pervasive computing deals with linking changes in the environment with computer systems such as RFID, GPS, ambient intelligence and other emergent context-aware applications [79]. In criminal cases, psychological profiling has given way to DNA matching. In consumer products, commodity logistics have given way to RFID databases. Genomics are the universal identification of life abstract; biometrics is considered as the universal identification of life in particular; collaborative filters are the universal identification of life in the relational [80]. Biometrics is specified as the science of recognizing an individual based on psychological or behavioral traits. Biometric systems, which rely on the evidence of fingerprints, hand geometry, iris, retina, face, hand vein, facial thermo-gram, signature, voice, etc., are deployed as a means of establishing and validating identity [81]. Privacy issues related to digital identity would inevitably rise as far as coincidence between happening and storage becomes more persistent in the future.

2.6.6 *Frauds, Misuse, Fake Profile and Crimes of Identity*

Identity fraud is a profitable enterprise. "Individuals have an asset called their identity", said Dr Tom Ilube, CEO of a data security company. He adds: "it is valuable to you and valuable to those people that want to abuse it" [82]. Fraud is rising rapidly because people are posting personal facts on the Web as well as government agencies are steadily making databases available online. These databases include birth, marriage and death certificates, credit histories, voter registrations and property deeds [83]. Security, identity theft, incorrect computer records, credit rating destruction, privacy, online purchasing and banking, loss of identity, misuse of personal information, phishing, identity cards, behavioral monitoring and tracking, the list of concerns goes on and on [84]. The Liberty Identity Theft Task Group, defined the three stages of identity theft as: (1) stealing identity data: while the numbers and stories about identity data loss are sensational, companies that suffer this traditionally only faced embarrassment and a bruised reputation; (2) hijacking existing accounts: 80 % of phishing attacks are against financial services[18]; and (3) concocting new accounts: the fraudulent opening of new accounts using another's identity is more dangerous because valid credentials are given to the criminal. When identity credentials are given to the wrong person, the strength of identity technology is powerless to help [85]. In addition, identity theft and fraud rate is increased due to risks posed by data deluge. Data deluge poses risks such as disks full of social-security data go missing; laptops loaded with tax records left in taxis; credit-card

[18] Anti-phishing working group: http://www.antiphishing.org.

numbers are stolen from online retailers; and Big Brotherishness of customers' personal information. The consequence is privacy breaches, identity theft and fraud [85]. More than 800 million active users in Facebook, around half of them currently access Facebook through their mobile devices [86]. As far as social networks are attracting more people, digital identity within these networks becomes more fragile and easily fall prey to social engineering traps. The case of Robin Sage experiment [87, 88] illustrate how fragile is digital identity in social networks and how it is easy to create a fake online profile that refers to nonexistent offline person. The Robin Sage experiment was conducted by Thomas Ryan. He created blatantly false identity of a woman claiming to work for in military intelligence and then enrolling on various social networking websites. Ryan deliberately chose an attractive young female's picture to prove that appearance is crucial in trust and people's eagerness to connect with. After a month, Robin has accumulated connections to around 300 online social networks. Contacts included an array of executives at government entities, employees of global 500 corporations and throughout the experiment Robin was offered gifts, government and corporate jobs, and opportunities to speak at various security conferences. Ryan tried to highlight how easily trust is given in these spaces and how much different information gets leaked out through various networks. He recommends social network users to accept only contacts that they know or make a research on people before accepting contacts' requests. See more cyber-criminality for black-markets report [89].

2.6.7 Digital Identity Aggregation Drivers and Issues

We ascribe "Out of Many, One" from "E Pluribus Unum" [90], which is used in the Great Seal of the United States [91], to underline the idea behind the scene of digital identity aggregation and fusion. Profiles are either unified into one all-encompassing digital dossier or relationships are defined among them to form a single digital identity. Moreover, we use the expression to point out high and urgent societies' expectations and needs for digital identity fusion capabilities that help investigators to identify a terrorist blended in with many people.

Data fusion can drive organizations to make better use of the data they own and provide convenience by creating an information resource that is more powerful, more flexible and more accurate than any of the original data sources [92]. Early in the mid-1800s, Matthew Fontaine Maury of the American navy had the idea of aggregating nautical logs from ships crossing the Pacific to find the routes that offered the best winds and currents. He created an early variant of a "viral" social network, rewarding captains who submitted their logbooks with a copy of his maps. But the process was at that time very slow and laborious [93]. Las Vegas casinos have been pioneers in fusing data from various sources because they face so many schemes to rip them off. Watching Hollywood films such as Enemy of the State and the Jason Bourne trilogy shows that shadowy organizations have instant and easy access to all the databases for various security purpose, particularly to identify terrorists. DARPA researchers argued that the World Trade Center bombing of 1993 and the Oklahoma City bombing of 1995 might have been prevented if US public security services could have linked commercial databases to identify large purchases of fertilizer by non-farmers [92]. In addition, the author of the Economist 'Data Deluge' article [94] explains the current situation of digital identity aggregation and fusion by pointing out that despite years of large-scale efforts, law-enforcement and intelligence agencies' databases are still not effectively linked yet. He gives the examples of health care industry in which computerizing health records tend to run into bureaucratic, technical and ethical problems. The digitization of health records could have been helpful to spot and monitor health trends and evaluate the effectiveness of different treatments. We point out that features and tools offered by Naymz [63] such identity aggregator, reputation assessment tool, and reputation score 'RepScore' could inevitably help to build trust-based professional community. After 9/11, the American Defense Department launched a program called "Total Information Awareness" to compile as many data as possible: e-mails, phone calls, web searches, shopping transactions, bank records, medical files, travel history and much more. In his article [92] titled "Information of the World, UNITE!" published in Scientific American Magazine, Simson L. Garfinkel explains through a hands-on, real-life experience motivations of digital identity aggregation or fusion. He says: "A few years ago I bought a latte at Starbucks on the way to the airport, parked my car and got on a flight for the U.K. 8 h later I got off at Heathrow, bought a prepay chip for my cell phone and went to buy a ticket for the train into London, when my credit card gave up the ghost and refused to work anymore. Not until I got back to the U.S. did I find out what had happened. Apparently, the small purchase at Starbucks, followed by the overseas purchase of the cell phone card, had tripped some kind of antifraud data-mining algorithm in my credit-card company's computer. It tried to call me, got my voice mail and proceeded to blacklist my credit card. What I found so exasperating about the entire experience was that the computer should have known that the person using my card in England was me. After all, I had bought my plane ticket with that same card and had flown with a major U.S. carrier. Aren't all those databases supposed to be tied together?" [92]. In the next sections, we explain that mashing digital identity attributes, from credit-card bills to cell phone logs, poses technical, economic, legal and ethical problems. Below, motivations for security purpose are listed and explained.

2.6.8 Digital Identity Aggregation for Security Use Cases

A digital identity silos consolidation is considered as one of the current challenges and a critical step to secure access to information systems' assets [27]. Digital identity aggregation, synonym of 'digital identity silos consolidation', establishes relationship between distributed attributes. We use the term 'silos' to convey that digital identity attributes are rarely stored in one place but rather in diverse and various stores residing within multiple information systems. As a consequence, the individual is in one-to-many relationship with his identity. Merriam's dictionary defines 'to consolidate' in the meaning of to strengthen and to unite. Several use cases explain and illustrate the need of digital identity aggregation for security purpose. We detail three of the use cases: (1) applications and services may require more attributes to authorize the user accessing resources. This is reflected in the real world as a person, who is asked to provide more than one identity proof comprising different identity information to get a customized service. For instance, a customer is asked to provide a credit card and fidelity saving card in a movie store to take advantage of DVD prices rebates. Moreover, to get into some mistrusted or restrictive environments, such as national security organizations, a visitor is asked to provide more than one identity card; (2) provisioning an employee who leaves. Consolidating employee identity attributes across information systems and synchronizing them would allow recognizing the validity of his authentication performed inside and outside the information system; (3) online reputation systems are in use to trust parties and conduct secure online business. For instance, eBay reputation mechanism unifies member's transaction feedback history to calculate community members' reputations in the form of colored and shooting stars. In addition, we need not only just a consolidation but an effective attributes because a poor administration and maintenance of duplicated, out-of-date, and low-quality identity attributes may expose enterprise assets and resources at a high risk. From the subject and service provider perspectives, digital identity aggregation becomes a highly used tool to reduce identity theft. Currently, services providers are using advanced tactics, collectively known as identity scoring that allows monitoring online data mining, pattern recognition, even semantic analysis of information about a subscriber that appears on Web pages. Examples of firms that offer such services are Garlik [95] in England and MyPublicInfo [96] in U.S. Garlik offer 'data patrol' service to British residents by combing credit reports, public databases and Web sites for information about customers and presents them with a detailed profile. The profile should show whether criminals may be trying to use their personal facts to apply for credit cards, take out a loan, or register a fake driver's license or marriage certificate. MyPublicInfo pieces together a customer's 'public identity profile' for $79.95 and alert him or her to dubious changes for $4.95 a month [83]. Moreover, the subject must be able to combine selected claims made about himself by more than one identity authority into a minimal composite set of claims and be able to present them to relaying party, who could not be able to repudiate the original claims [59].

Many participants have different profiles within multiple social networks. From the user perspective, aggregating profiles would (1) increase convenience of the social experience: the participant can post a message to multiple friends within different social networks; (2) ease access control (identification, authentication, authorization, and accountability); and (3) attributes management. From the organizational perspective, social profiles aggregation would ease (1) participant's reputation management: HR department might aggregate a candidate's social profiles in order to decide whether to hire him or to reject his application. Another example is a student, who wants to know more about his professor, would make a Google search and professor's social profiles aggregation; and

(2) service personalization (profiling): in order to increase market shares, companies might aggregate client's social profiles to know more about their preferences and goals, as a consequence, they can personalize products and services. They might also consider the friends list of a client or business partner as prospect clients.

2.6.9 Economy of Digital Identity Aggregation: Digital Gold Mine

Today, organizations strive to capture and aggregate digital identities because they are convinced that is the new form of 'rué-vers-l'or'. Such agitation is comparable to the one that is used to be with hundreds of people when searching for gold, panning in the streams and digging mines. 'Gold Rush' (1925), the Charlie Chaplin's movie, is a true illustration of major gold rushes that took place in the nineteenth century in Australia, Brazil, Canada, South Africa, and the United States [76, 97, 98].

Digital identity attributes become publicly available and easy to access. Each person now leaves in cyberspace an increasingly amount of digital footprint when aggregated and unified, contributes to the definition of the subject's digital identity. Visible or invisible, left consciously or not, this set of data can be collected from various sources. The very first digital records of pre-natal scans could be shared on flicker and the obituary information on the Social Security Death Index (SSDI),[19] Find a Grave,[20] and Interment.net.[21] It happens also that other data could be available and collected through the one collected by diverse agencies and organizations on our behalf during our life, the blogs that are kept, the emails sent and the internet searches performed [66–68]. Maintaining and editing personal information in learning digital portfolio or personal profile within social network is much feasible and easier than the personal profile that is carried out kept by an employer 'googling' prospective employee, tracking activities as a citizen, and possibly inferring health problems from the visible activities in self-advocacy online groups. For instance, We Feel Fine [99], Fig. 2.7, is a people feeling aggregation engine that harvests automatically human feelings from a large number of blogs every 10 min. Compiled blog data [100] comes from a variety of online sources, including LiveJournal, MSN Spaces, MySpace, Blogger, Flickr, Technorati, Feedster, Ice Rocket, and Google. The engine scans blog posts for occurrences of the text fragments 'I feel' and 'I am feeling'. The approach was inspired by techniques used in Listening Post project [101].

The value of digital identity increases as much as substantial quantity of digital identity attributes has been collected and aggregated. Many people search engines are evolving to better provide services by aggregating people digital identity

[19] http://ssdi.rootsweb.ancestry.com/

[20] http://www.findagrave.com/

[21] http://www.interment.net/

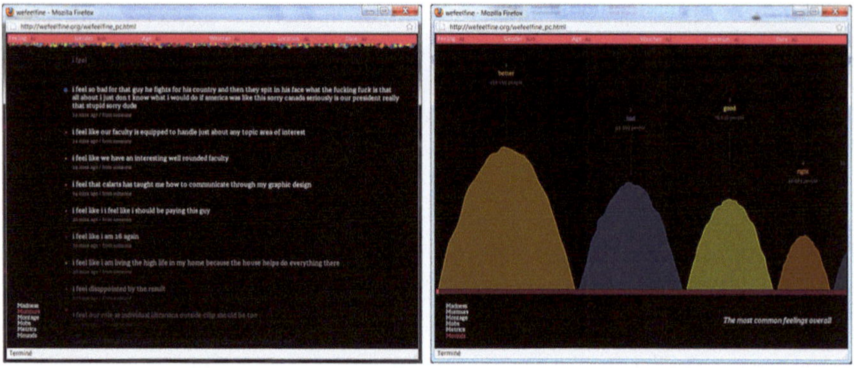

Fig. 2.7 People feelings murmurs and emotions aggregator

attributes. 123People[22] engine provides information on people to learn more about a person, an acquaintance, a colleague, a potential collaborator. Having its roots in Austria, 123people aggregates information of the digital identity of a person on the Web taken from multiple sources such as Web pages, social networks, images, videos, blogs, micro-blogging platforms, and emails [102]. Others such as Spock[23] and USsearch[24] are providers of people search and background checks that work jointly to provide free aggregated digital identity information and paid service to access on sensitive and premium information such as criminal record. Another service that provides obituary information is SSDI Index. The person enters the first and last name, then the SSDI Index resource turns up full name, birth and death dates, last known residence, last benefit, social security number, and state in which the social security card was issued. Other record-related information is available upon order. As an example, we use the Social Security Death Index (SSDI) service provider to look for 'Abraham Lincoln' personal information in US public registries. The result is presented in the following screenshot, Fig. 2.8.

Another example of public records aggregator and people finder is Intelius.[25] The system reports genealogy records that comprise phone numbers, address history, birth certificates, death records, marriage licenses and divorce decree. It allows tracing family tree by saving, adding, and joining records together. Moreover, the system provides neighborhood and property information such as home value, sales history, property details and ownership information. In Fig. 2.9, Intelius shows Ghazi Ben Ayed's public record as he was a resident of Milwaukee, WI from 1998 to 2000. It makes public personal data such as his mother's full name in the relative column: 'Zahra Ben Ayed'. When the user heats the View

[22] www.123poeple.com

[23] www.spock.com

[24] www.ussearch.com

[25] http://www.intelius.com/

Viewing 1-20 of 21 — 1 2 | Next

Name	Birth	Death	Last Residence	Last Benefit	SSN	Issued	Tools	Order Record?
ABRAHAM LINCOLN	17 Sep 1887	Dec 1966	32713 (Debary, Volusia, FL)	(none specified)	001-12-3083	New Hampshire	SS-5 Letter Add Post-em Search Ancestry.com	
ABRAHAM LINCOLN	22 May 1895	Feb 1970	15108 (Coraopolis, Allegheny, PA)	(none specified)	069-03-6847	New York	SS-5 Letter Add Post-em Search Ancestry.com	
ABRAHAM LINCOLN	22 Jun 1911	Aug 1983	11758 (Massapequa, Nassau, NY)	(none specified)	089-12-8975	New York	SS-5 Letter Add Post-em Search Ancestry.com	
ABRAHAM LINCOLN	23 Jan 1907	Nov 1974	12941 (Jay, Essex, NY)	(none specified)	116-01-1374	New York	SS-5 Letter Add Post-em Search Ancestry.com	
ABRAHAM LINCOLN	15 Nov 1926	May 1979	(not specified)	14729 (East Otto, Cattaraugus, NY)	117-14-3679	New York	SS-5 Letter Add Post-em Search Ancestry.com	
ABRAHAM LINCOLN	24 Nov 1889	15 Nov 1966 (V)	19146 (Philadelphia, Philadelphia, PA)	(none specified)	172-24-6105	Pennsylvania	SS-5 Letter Add Post-em Search Ancestry.com	
ABRAHAM LINCOLN	13 Nov 1900	04 Aug 1990 (V)	17404 (York, York, PA)	(none specified)	178-16-3226	Pennsylvania	SS-5 Letter Add Post-em Search Ancestry.com	
ABRAHAM S LINCOLN	27 Jan 1919	15 Mar 2005 (V)	79925 (El Paso, El Paso, TX)	(none specified)	225-01-5120	Virginia	SS-5 Letter Add Post-em Search Ancestry.com	
ABRAHAM ELIJAH LINCOLN	08 Apr 1918	26 Aug 2007 (P)	29212 (Columbia, Lexington, SC)	(none specified)	250-09-9182	South Carolina	SS-5 Letter Add Post-em	

Terminé

Fig. 2.8 Abraham Lincoln obituary information in US public records

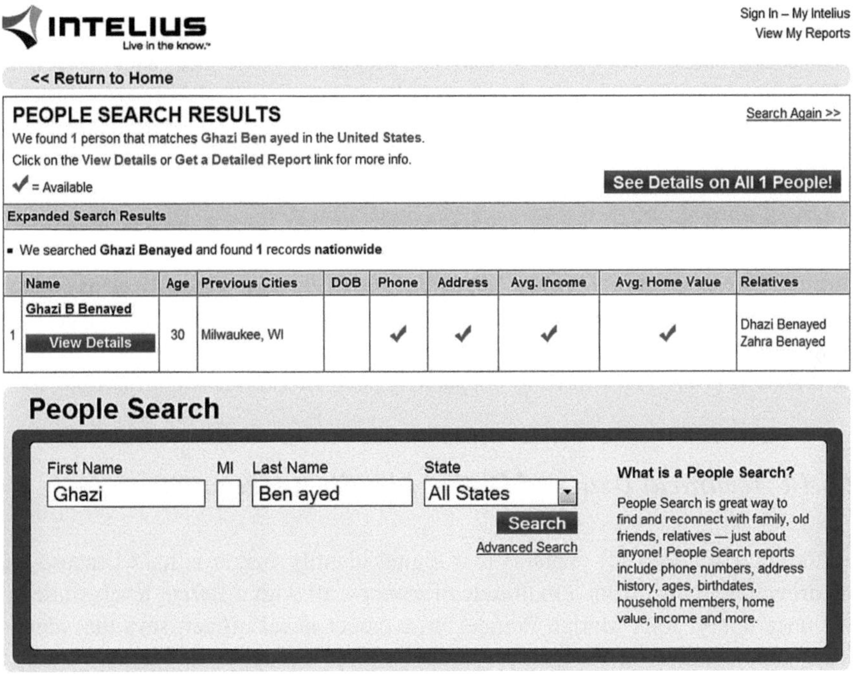

Fig. 2.9 Result of 'Ghazi Ben Ayed' searching in US public record

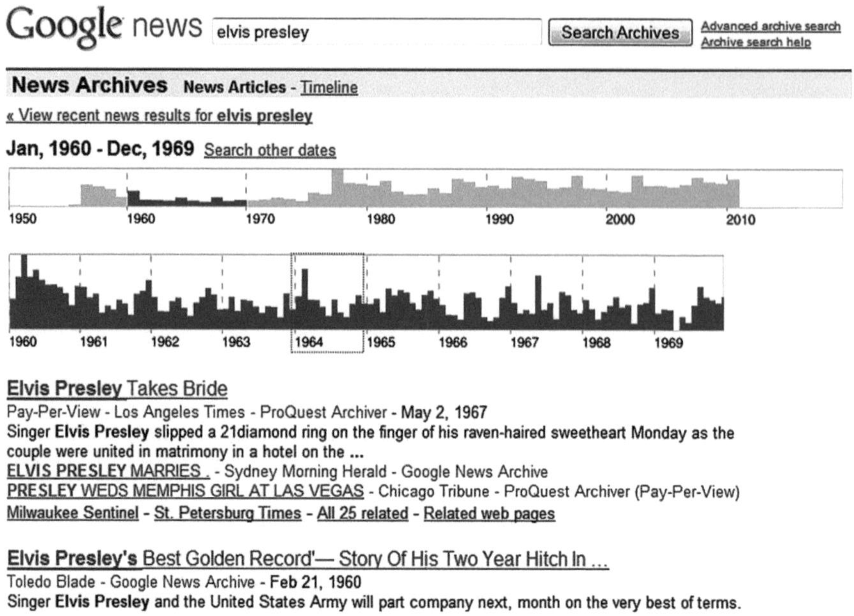

Fig. 2.10 Elvis Presley archives available through google news search archives

Details red button requesting to edit the public record-related data located at Wisconsin authorities, the system asks to order and pay, at special or regular prices.

In addition, users of Google news archive search[26] can explore historical archives about events, people or ideas and see how they have been described over time. In addition, users can also see a historical overview of the results by browsing an automatically generated timeline. Search results include content from a number of sources, through content digitized by Google and online archival materials that Google crawled. Search results can include content that is freely accessible as well as content that requires a fee. Articles related to a single story or person within a given time period are grouped together to allow users to see a broad perspective on the topics they are searching [103]. Figure 2.10 shows publicly published Elvis Presley information.

2.6.10 Technical Issues of Digital Identity Aggregation

In 2008, the author [92] explains that digital identity fusion is hard because we are drowning in data from a multitude of sources, all with different levels of detail and uncertainty. John Marlan Poindexter, a career naval officer, says that identifying the signatures of terrorist preparations in an ocean of data is much harder

[26] http://news.google.com/archivesearch

than finding subs in an ocean of water. In addition, Poindexter argues that oceans may be huge but every spot can be uniquely identified by a latitude, longitude and depth. However, data oceans are not so easily to be categorized. Much of information are spread across millions of individual computer systems and hidden to the authorities. In addition, oceans are not doubling in size every few years like data oceans. Major issues are: (1) data quality. Much of the personal data in databases may not be accurate and they are riddled with errors and meaningless coincidences. A Scientific American editor ordered an US $80 report from an online consolidator of digital identity, including criminal, real-estate and bankruptcy records. It was riddled with errors such as misspellings and confusion with namesakes. The report showed no signs of identity theft! Currently, new algorithms overcome only some of these hurdles but not all of them; (2) making sense (semantics) of data fusion. Users are sometimes unaware of the digital bread crumbs they leave but companies are increasingly linking isolated databases together into one data scheme could infect a person's entire digital identity and reputation either by stealing data scheme or through attributes aggregation bias, particularly decontextualization of digital identity by data mining algorithms. Yet another problem for data fusers is; (3) identity resolution, which is matching up the various names and account numbers with the right individual by taking into account cultural variation in names and other business-related rules [92, 104]. In online world there may be dozens of people sharing the same name and dozens of names used by the same person, thus the issue deals with ontology and syntax of digital identity attributes. Person's first name may be listed in one database as Robert, in another as Rob and in a third as Bob. A person whose Arabic name is Haj Imhemed Otmane Abderaqib in West Africa might be known as Hajj Mohamed Uthman Abd Al Ragib in Iraq. Casinos have funded development of a technique called NOnobvious Relationship Analysis (NORA), which combines identity aggregation and resolution with databases of credit companies, public records and hotel stays [92].

Figure 2.11 sums up digital identity aggregation technical issues and illustrates that attributes semantics, ontology, syntax and interoperability issues arise whenever and authority needs to aggregate a multiple digital identity attributes in order to decide whether to provide a service to the subject. For example, how computers could recognize that the short names 'G. Ben Ayed' and 'Ghazi B. Ayed' are referring to the same person with a full name 'Ghazi Ben Ayed'? In addition, names written with typo errors such as 'Gazi Benayed' and 'Ghasi Bennayed', the ones written in other languages and following cultural semantics such as Hispanic, Japanese, Chinese and Arabic, or Arab names written with Latin font could be automatically recognized as being part of the same person's identity? The authors [31] explain that identity management service must support vocabulary definitions of identity attributes. A fundamental assumption is that all parties concerned with identity services share a common ontology and semantic web metadata formats such as Resource Description Format (RDF) and RDF Schema.

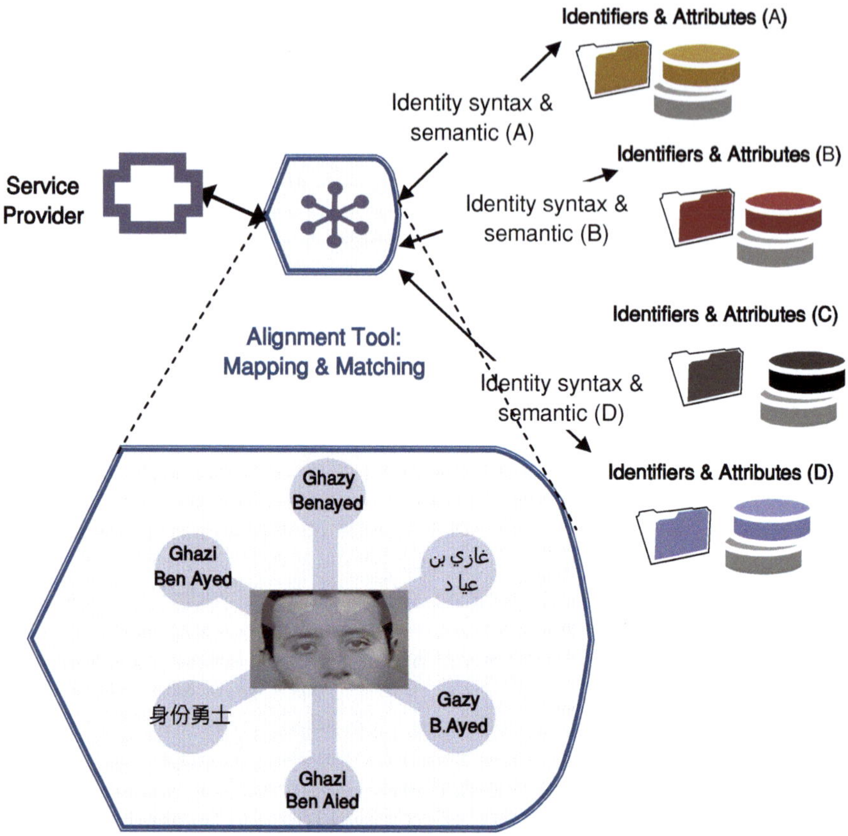

Fig. 2.11 Digital identity aggregation technical challenges

2.6.11 Digital Identity Aggregation Systems and Algorithms

Algorithms of data fusion can trace its heritage back to the computerized matching programs of 1970s. US government authorized the creation of the Federal Parent Locator Service that denies a wide range of federal benefits to parents who are behind on their child support. Those data are fused with digital identity of recently employed parents who are not up to date on their payments so that their wages can be garnished [92]. After 9/11, the American Defense Department launched a program called "Total Information Awareness" to compile as many data as possible: e-mails, phone calls, web searches, shopping transactions, bank records, medical files, travel history and much more [105].

The program works by building hypotheses based on existing profiles and then revising these hypotheses as other digital identity attributes become available. In the 1990s software engineer Jeff Jonas developed a system that could match the names in a casino's computers with other sources of information. Figure 2.12, which is adapted

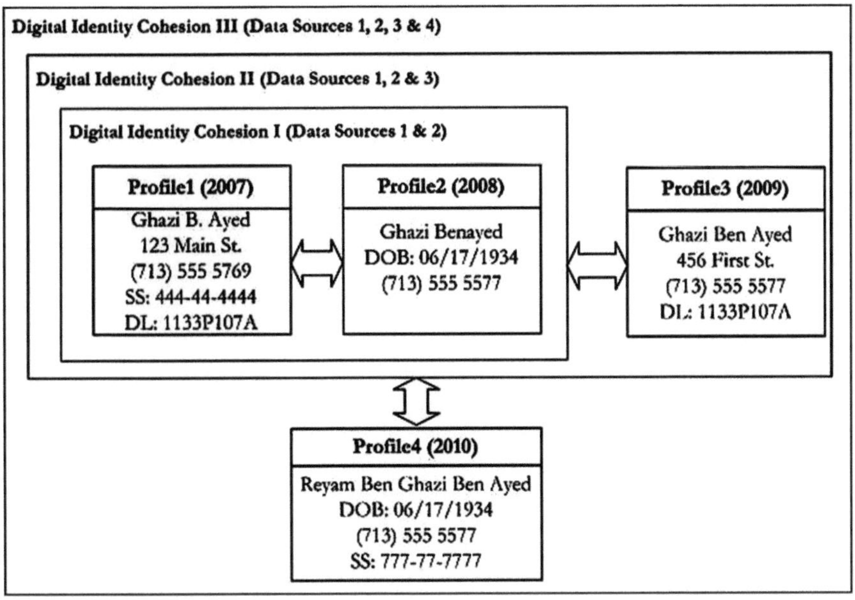

Fig. 2.12 Casino's digital identity fusion algorithm developed by Jeff Jonas in 1990s

from [92], shows that four of the profiles reside in different locations and have been collected in different periods of time. Digital identity aggregation I combines profile1 and profile2 and each of them holds different attributes, so the system provisionally assumes they represent different individuals. In aggregation II, the system infers that profile3 holds attributes common to both previous profiles: the driver's license number from one and phone number from the other. So the system reassigns all three to the same individual. Finally, digital identity aggregation III shows that profile4 includes a birth date matching with profile2, thus, the system deduces that the four profiles actually represent two individuals. The program guesses that the two may be father and son since they share the same surname and phone number. In 2005, Jonas sold the system and his company to IBM, which has added a feature called anonymous resolution. Two organizations can determine whether they share the digital identity of an individual in their databases without revealing digital identities of all people who do not match. The technique works by comparing cryptographic hashes instead of digital identity attributes. Currently, most algorithms of data fusion have some kind of sensitivity adjustment. Tipping the scale to the right, and the system fails to find genuine matches; tipping it to the left, the system turns out to be wrong because too many predictions are achieved. Another important issue raised by data aggregation is to find an algorithm that it never confuses original data with a conclusion inferred from those data [92].

Economic gains should justify fusion costs. In 1994, Roger Clarke of the Australian National University in Canberra studied computerized matching programs maintained by federal and state governments in the U.S. and Australia. These systems

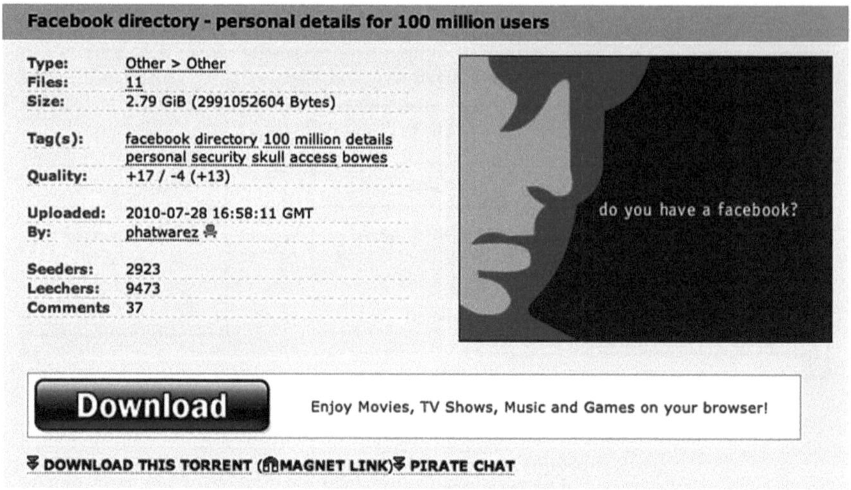

Fig. 2.13 A ready-to-download file comprising details of 100 million Facebook users

scanned millions of records and flagged thousands of potential "hits." But most of them turned out to be false positives. The benefits did not justify the costs of collecting data, training personnel and chasing down the false positives. The same author argues that many people feel that if a data-fusion program could anticipate and stop a major terrorist attack, it would be worth whatever it cost [92]. However, from the ethical and legal perspectives, linking together databases into a single profile through the process of data fusion is still the bête noire of privacy advocates. They advocates still considering that identity data aggregators use personal information for purposes other than the ones for which it was originally acquired [92]. The author of How To Be Invisible book [89] states: "Do not, as long as you live, ever again allow your real name to be coupled with your home address". This is to point out that preserving privacy is a matter of conscience. Privacy issues are detailed in Chap. 4.

Many use cases illustrate the dangers of maintaining digital identities at a poor security level. Almost 3 GB file that contains 100 million Facebook users has been made available on a torrent site downloadable by absolutely anybody in July 2010 see Fig. 2.13. Ron Bowes of Skull Security created a script [106] that harvested user information from Facebook's user directory [107].

Ron's idea was to spider and generate first-initial-last-name list and once he had the name and URL of a user, he aggregated users' pictures, friends, and information about them, with some other details. He wrote a Ruby script to download the full Facebook users' directory and link personal details to the corresponding first, last, and usernames. The results were 171 million names (100 million unique) [107].

The file, Fig. 2.13, contains the URL of every searchable Facebook user's profile, the name of every searchable Facebook user, both unique and by count, and processed lists, including first names with count, last names with count, and potential

Table 2.1 Top Facebook usernames' lists

A first initial and last name-based list	A first name and last initial-based list	A first name dot last name-based list	A first name-based list	A last name-based list
129369 jsmith	100225 johns	17204 john.smith	977014 michael	913465 smith
79365 ssmith	97676 johnm	7440 david.smith	963693 john	571819 johnson
77713 skhan	97310 michaelm	7200 michael.smith	924816 david	512312 jones
75561 msmith	93386 michaels	6784 chris.smith	819879 chris	503266 williams
74575 skumar	88978 davids	6371 mike.smith	640957 mike	471390 brown
72467 csmith	85481 michaelb	6149 arun.kumar	602088 james	386764 lee
71791 asmith	84824 davidm	5980 james.smith	584438 mark	360010 khan
67786 jjohnson	82677 davidb	5939 amit.kumar	515686 jason	355639 singh
66693 dsmith	81500 johnb	5926 imran.khan	503658 robert	343220 kumar
66431 akhan	77800 michaelc	5861 jason.smith	484403 jessica	324972 miller

usernames with count, as presented in Table 2.1 [107, 108]. Even if the user opts out of inclusion in the search, he could still appear on the directory page of a searchable friend. The statistical lists don't pose any security threat to Facebook users; however, data could be useful for building automated account cracking software. Lists of the most common names can be used to assemble a good dictionary of potentially popular usernames for use in tools that attempt to identify and crack user accounts [109].

2.6.12 Digital Native's Perception of Identity

What is the impact of digital identity and privacy on the "digital native"? The term "digital native" means the generation that grew up with Internet and new information technologies. "Digital Natives" were born after 1980, when social digital technologies came online. They carry mobile devices all times not just to make phone calls but also to send text messages, surf the Internet, and download music. They've been living with mobility, speed access to information, learning with media, participatory action, co-creation of value, etc. [110–112] In the shadow of the daily growth of the global population in general and particularly the digital native one [111, 113], digital identity would play major role in the next few years. Digital Natives live much of their lives comfortably online, without thinking of their digital identity and their real-space identity as separate things. They just have an identity, which is a representation in two, or three, or more different spaces. Digital Natives are constantly connected. Even as they sleep, connections are made online, in the background; they wake up to find them each day. They connect to social networks, IM, and share photos with friends all over the world. Digital Natives are creating parallel worlds on sites like Second Life. And after they do, they record parts of that world and post a video of it on YouTube or Daily Motion in a 'machinima' art form. Digital Natives perceive information as something they can control and reshape in new and interesting ways. They edit a profile on MySpace or encyclopedia entries on Wikipedia, make a movie or online video, or download a hot music track—whether lawfully or not. Digital Natives can rework media, using off-the-shelf computer programs. Research means Google search and particularly Wikipedia before diving deeper into a topic. Most Digital Natives don't buy newspapers ever but they get it in new ways and in a wide variety of formats. In the process of spending so much time in this digitally connected environment, Digital Natives are leaving more traces of themselves in public places online. With every hour they log online, they are leaving more tracks for marketers—and pedophiles, for that matter—to follow. Digital Natives' ideas about privacy, for instance, are different from those of their parents and grandparents but how? The repercussions of these changes in the near future will be profound for all of us. The Digital Natives has global culture in scope and nature whether physically based in different cities, countries and continents [111].

2.6.13 Issues and Concerns Associated with Handling the Digital Afterlife

In his article preparing for the digital afterlife [114], Duncan Jefferies questions how should we handle digital legacy? How should we deal with online accounts such as Facebook and PayPal logs off for good? It might depend on the law, but by default digital assets are "the property of the estate, even if they're property with no value". Some assets, such as blogs and photographs, may also be subject to intellectual property law. "People aren't very aware of what you might call their living online legacy—potential employers looking at their Facebook accounts, for example. The issue of what happens to that information after their death is an extension of that" says Yorick Wilks, a senior research fellow at the Oxford Internet Institute. Facebook puts the profile of deceased person into a memorial state upon notification of their death. Their status is removed, they are withdrawn from any groups and access is set to "friends only". Couldn't his descent being part of his social circle of friends? Donna Rawling lost her husband and she says: "I managed to wrap up his affairs, but the area that I was left with was his presence on the web". Several companies aim to help people to better handle digital legacy by providing Digital deposit accounts playing the role of "electronic safe deposit box", where people can easily upload login details for digital assets and specify who will receive them posthumously. Examples are LegacyLocker,[27] SlightlyMorbid[28] and Deathswitch.[29] Deathswitch provides an automatically

[27] http://legacylocker.com

[28] https://www.slightlymorbid.com

[29] http://www.deathswitch.com

prompts people for their password on a regular basis. If nothing is received after several prompts, the system deduces that the user is already dead or critically disabled, thus, messages are sent to pre-selected recipients. As they are large repositories of passwords, does the hacking community perceive these systems as a virtual El-Dorado? Could these systems not expire before its customers do? "People aren't very aware of what you might call their living online legacy—potential employers looking at their Facebook accounts, for example. The issue of what happens to that information after their death is an extension of that", says Yorick Wilks, a senior research fellow at the Oxford Internet Institute [114].

2.6.14 Digital Identity, Online Reputation and Metadata

As data become more abundant, the main problem is no longer finding the information but accessing it easily and quickly. What is needed is metadata, which is information about information, to organize the cornucopia of information provided by the internet. In Assyria around three millennia ago clay tablets had small clay labels attached to them to make them easier to tell apart when they were filed in baskets or on shelves. The idea survived into the 20th century in the shape of the little catalogue cards librarians used to note down a book's title, author, subject, and so on before the records were moved onto computers. The actual books constituted the data, the catalogue cards the metadata. Bar coded and RFID package labels are other examples of metadata. Today, metadata are undergoing a virtual renaissance since many companies are using it to organize information. Google's search engine creates PageRank metadata to organize web pages by structuring the information, ranking it in order of its relevance to the query. Google handles around half the world's internet searches, answering around 35,000 queries every second. Metadata are a potentially lucrative business. "If you can control the pathways and means of finding information, you can extract rents from subsequent levels of producers," explains Eli Noam, a telecoms economist at New York's Columbia Business School [115].

Metadata could directly affect the digital identity and online reputation since metadata are increasingly become available on the net. Photos uploaded to the website Flickr contain metadata such as when and often where they were taken, as well as the camera model, which could be useful for future buyers. But with the advent of Web 2.0, internet users tag web sites, documents, photos and videos helping to label unstructured information so it can be easily found through folk-minds such as Delicious,[30] Diigo,[31] and Technorati.[32] For any reason, such as for having fun or creating a buzz on the net, Internet users could also instead labeling

[30] http://www.delicious.com

[31] http://www.diigo.com

[32] http://technorati.com

a photograph of Barack Obama as "president", they might bookmark it "sexual harassment". Thus, this phenomenon would have a negative side affecting the people's digital identities and reputations [115].

2.6.15 Digital Identity Issue with Cyborg Enhancement

Identity and privacy issues are immediately important with enhancing implant technology, even in the case of relatively straightforward identification devices. A 'Cyborg' is a cybernetic organism, part human, part machine, and is formed by the direct connection between human and technology. In 2002, an implant experiment was carried out through an online collaboration between Columbia University and Reading University. It consists of linking the nervous system of a human with the internet. Intents and purposes the body of that individual does not stop as is usual, but rather extends as far as the Internet takes it. In this case, the human brain was able to directly control a robot hand on a different continent-the Cyborg body extended across the Atlantic Ocean. In this respect, by linking the mental functioning of a human and machine, a hybrid identity is created. By connecting the human nervous system with technology, this not only affects the nature of an individual's identity but also raising questions as to a new meaning for 'I'. Who are we if our brain/nervous system is part human part machine? Privacy issues are also pertinent when considering signals being sent into and out of the brain. Feelings, emotions, and even inter-thoughts could potentially be modified by electronic signals alone. Network hacking is far more serious if your brain is permanently connected into the network. Software viruses and biological viruses become, effectively, the same thing. Hence, security, screening and anti-viruses take on much more importance [116].

2.6.16 Digital Identity in Big Data Era

Information has gone from scarce to superabundant and the quantity of information in the world is soaring and becomes astronomic. Joe Hellerstein, a computer scientist at the University of California in Berkeley, calls it "the industrial revolution of data". Scientists and computer engineers have coined a new term for the phenomenon: "big data" [94, 117]. Authors provide many examples to illustrate the importance of data deluge. Headquartered in Hong Kong, Li and Fung Ltd.,[33] a major global distribution service company, saw during 2008 one hundred gigabytes of information flow through its network each day; but today the amount has increased tenfold. During 2009, US army's aircraft flying over Iraq and Afghanistan sent back around 24 years' worth of video footage. The same author predicted that new aircraft models that

[33] http://www.lifunggroup.com/front.html

are being deployed in 2010 will produce ten times as many data streams as their predecessors, and those in 2011 will produce 30 times as many. He adds that according to one estimate, mankind created 150 EB (billion gigabytes) of data in 2005 and it will create 1,200 EB in 2010 [94]. In 2000, the telescope of Sloan Digital Sky Survey SDSS collected more data in the first few weeks than had been collected in the entire history of astronomy. Today's SDSS archive contains a whopping 140 TB (240 bytes) of information. A new generation of telescope, the Large Synoptic Survey Telescope, will acquire that quantity of data every 5 days. The retail giant Wal-Mart handles more than one million customer transactions every hour, feeding databases estimated at more than 2.5 PB—the equivalent of 167 times the books in America's Library of Congress [117]. Photobucket, an online photo-sharing service, claims to host more than 4.7 billion digital photographs as of 2008. Facebook reports more than 3 billion photographs, less than 4 years into its existence [111], and reached to home 40 billion photos in 2008 [117]. The author of the Economist article [93] explains that the amount of information is growing at a terrific rate. He adds that experiments at the Large Hadron Collider at CERN generate 40 TB every second and, in 2008, U.S. households were bombarded with 34 gigabytes per person per day [93]. YouTube manages video uploads of 5 h/min early in 2007 to more than 35 h/min in 2010 [118]. The author points that several reasons are driving digital information explosion. He estimates that amount of information increases tenfold every 5 years for the following main reasons: (1) technology is the obvious one. Digital devices soar such as sensors and gadgets are digitizing lots of information that was previously unavailable; (2) there are now many more people who interact with information. Between 1990 and 2005 more than 1 billion people worldwide entered the middle class. As they get richer they become more literate, which fuels information growth [117].

Companies could prosper by gasping new opportunities around big data. The author says that companies could 'pluck the diamond from the waste' by exploiting big data opportunities. Analyzing data could help to spot business trends, prevent diseases, and combat crime. Effective data management could unlock new sources of economic value, provide fresh insights into science and hold governments to account. For instance, exploiting and mining crime figures, maps, details of contracts and statistics that public services are putting into the public domain, or provide the tools for others to do so [94, 117]. Many businesses are providing services based on the access to government data, which recently are made available online. The state is a big generator, collector and user of data. It keeps records on every birth, marriage and death, compiles figures on all aspects of the economy and keeps statistics on licenses, laws and the weather. Until recently all these data have been locked tight and even if they were made publicly accessible they were hard to find, and aggregating lots of printed information is notoriously difficult. Today, things have changed "Government information is a form of infrastructure, no less important to our modern life than our roads, electrical grid or water systems," says Carl Malamud, the boss of Public.Resource Group[34] that puts government data online.

[34] http://public.resource.org

He was responsible for making the databases of America's Securities and Exchange Commission available on the web during 1990s [119]. The author of the "Clicking for Gold" article [120] explains that the trail of clicks that internet users leave behind from which value can be extracted is becoming a mainstay of the internet economy. "What we are seeing is the ability to have economies form around the data" says Craig Mundie, head of research and strategy at Microsoft. Data are becoming the new raw material of business. Farecast,[35] a part of Microsoft's search engine Bing, can advise customers whether to buy an airline ticket now or wait till the price to come down by analyzing 225 billion flight and price records. Amazon.com is not only tracks the books the user purchases, but also keeps a record of the ones the user only browses in order to recommend other books to him. Information that would be gathered from Amazon's e-book, the Kindle, is probably even richer, how long a user spends reading each page, whether he takes notes and so on [117, 120]. Business intelligence and analytics, which is performing statistical operations for forecasting or uncovering hidden correlations, may allow to firms to gain pay-offs by operating more efficiently, picking out trends and improving forecasting. "Torture the data long enough and they will confess to anything" is a humorous quip made by statisticians to encourage making the most of data. A few years ago business intelligence technologies were available only to big companies, but today the technology has moved into the mainstream. This is due to the fall of the price and better performance of hardware, software and storage. In addition, companies are collecting more data, which in the past they were kept in different systems that were unable to talk to each other, such as finance, human resources or customer management. Now the systems are being linked, and companies are using data-mining techniques to get, "a single version of the truth", which means a complete picture of their operations. Best Buy,[36] an international electronics retailer, found that 7 % of its customers accounted for 43 % of its sales, so it reorganized its stores to concentrate on those customers' needs. The author highlights that data torture depends the accuracy of the information that companies hold. In a study by IBM, half of the managers that are quizzed did not trust the information on which they had to make decisions. Currently, many businesses are increasingly moving to capture accurate data by analyzing real-time information flows instead of stored information about past transactions. Two technology trends are helping to fuel these new uses of data: cloud computing and open-source software. Cloud computing allows organizations to lease on-demand computing power, rather than having to acquire expensive equipment. A free programming language called R[37] lets companies examine and present big data sets, and free software called Hadoop[38] now allows ordinary PCs to analyze huge quantities of data that previously required a supercomputer.

[35] http://www.bing.com/travel

[36] http://www.bestbuy.com/site/index.jsp

[37] http://www.r-project.org

[38] http://hadoop.apache.org

Two major issues/difficulties that faces data deluge: (1) information storage capabilities: the current situation is a result of a rapid collection of data in a short time and an amount of data that exceeds the available storage space. Based on the forecast of IDC, in 2011, global information will reach around 1,750 EB and available storage about 800 EB. The flood of data from sensors, computers, research labs, cameras, phones and the like surpassed the capacity of storage technologies in 2007; (2) analysis and extraction capabilities of useful information: Alex Szalay, an astrophysicist at Johns Hopkins University, notes that the proliferation of data is making them increasingly inaccessible and he points that we should be able to make sense of them. Only few industries have developed such capabilities. Credit-card companies monitor every purchase and can identify fraudulent ones. They found that stolen credit cards are more likely to be used to buy hard liquor than wine for many reasons such as it is easier to fence. Insurance firms combine clues to spot suspicious claims. They found that fraudulent claims are more likely to be made on a Monday, since policyholders who stage accidents tend to assemble friends as false witnesses over the weekend. Mobile-phone operators, meanwhile, analyze subscribers' calling patterns to offer them customized attractive promotions. Also, retailers, offline as well as online, can tailor promotions to particular customers' preferences. The oil industry uses supercomputers to trawl seismic data before drilling wells [93, 94, 117]. In addition, another concern as the torrent of information increases is energy consumption. Processing huge amounts of data takes a lot of power. "In 2–3 years we will saturate the electric cables running into the building," says Alex Szalay at Johns Hopkins University. "The next challenge is how to do the same things as today, but with ten to one hundered times less power". The NSA in 2006 came close to exceeding its power supply, which would have blown out its electrical infrastructure. Both Google and Microsoft put some of their huge data centers next to hydroelectric plants to ensure access to enough energy and at a reasonable price [105].

Ensuring data security and protecting privacy is becoming harder as the information multiplies and is shared widely around the world. According to Cisco, by 2013, the amount of traffic flowing over the internet annually will reach 667 EB and the quantity of data continues to grow faster than the ability of the network to carry it all [117]. A researcher of the University of California in San Diego says: "information created by machines and used by other machines will probably grow faster than anything else". He adds that "this is primarily 'database to database' information—people are only tangentially involved in most of it" [93]. The author of the article "new rules for big data: regulators are having to rethink their brief" [121] points that current information flows in an era of abundant data are changing the relationship between technology and the role of the government. He adds that many of today's regulations are not brought up-to-date such as privacy laws, which they were not designed for networks, and rules for document retention presume paper records. Now information becomes interconnected and that's why nations are increasingly in need of global rules. The same author mentions that new information-related principles should cover the following broad areas: information privacy, security, retention, processing, ownership, and integrity to reduce risks posed by the age of big data sets [121]. More details are given by the author [94] to explain the consequences of only two data deluge risks, which are identity theft and fraud, and privacy breaches. He explains that they are consequences of stolen databases, such as disks full of social-security data are missed, laptops loaded with tax records are left in taxis, credit-card numbers are stolen from online retailers. Privacy infringements are encountered in daily basis. For instance, we can witness the periodic fusses when Facebook or Google unexpectedly change the privacy settings on their online social networks, causing members to reveal personal information unwittingly. A more sinister threat is encountered when governments compel companies to hand over personal information about their customers. In order to deal with the drawbacks of data deluge, the author suggests that people should have greater ownership, access, and control over their digital identity. For instance, Google allows users to see what information it holds about them, and lets them delete their search histories or modify the targeting of advertising. Secondly, organizations should be required to disclose details of security and privacy breaches to encourage managers to take information security and privacy more seriously. Finally, organizations should be subject to an annual digital identity and privacy audit to encourage organizations to keep their security measures up to date [94].

References

1. J. Amos, Etched ostrich eggs illustrate human sophistication (2010), Available: http://news.bbc.co.uk/2/hi/science/nature/8544332.stm. Accessed 29 Aug 2010
2. P.J. Texier, *Une découverte remet en question l'origine de "l'écriture"* (Science & Vie, 2010)
3. Wikipedia, Diepkloof Rock Shelter (2010), Available: http://en.wikipedia.org/wiki/Diepkloof_Rock_Shelter. Accessed 29 Aug 2010
4. P.J. Texier et al., A Howiesons Poort tradition of engraving ostrich eggshell containers dated to 60,000 years ago at Diepkloof Rock Shelter, South Africa, (2010), Available: http://www.pnas.org/content/107/14/6180.full.pdf+html. Accessed 29 Aug 2010
5. L. Casson, *Libraries in the Ancient World* (Yale University Press, 2002)
6. A. Whitaker, Nippur, Available: http://www.ancient-wisdom.co.uk/iraqnippur.htm. Accessed 25 Sept 2010
7. G.E.A. Akerlof, R.E. Kranton, *Identity Economics: How Our Identities Shape Our Work, Wages, and Well-Being* (Princeton University Press, 2010)
8. Merriam-Webster Online Dictionary, Definition of identity (2010), Available: http://www.merriam-webster.com/dictionary/identity. Accessed 20 Sept 2010
9. The American Heritage Dictionary of the English Language, Definition of identity (2009), Available: http://www.thefreedictionary.com/identity. Accessed 20 Sept 2010
10. H. Noonan, Identity, in *Stanford Encyclopedia of Philosophy* (2009)
11. Wikipedia, Nasreddin (2010), Available: http://en.wikipedia.org/wiki/Nasreddin. Accessed 23 Sept 2010
12. I. Shah, *The Exploits of the Incomparable Mulla Nasrudin—The Subtleties of the Inimitable Mulla Nasrudin* (Octagon Press, London, 1989)
13. C. Lindholm, *Culture and Identity—The History, Theory, and Practice of Psychological Anthropology* (Oneworld Publications, 2007)
14. E.T. Olson, Personal identity, in *Standford Encyclopedia of Philosophy* (2008)
15. Identity over time, in *Stanford Encyclopedia of Philosophy* (2005)

16. G. Group, Identity/identity formation, in *Gale Encyclopedia of Psychology* (2001)
17. Wikipedia, Identity (philosophy) (2010), Available: http://en.wikipedia.org/wiki/Numerical_identity#Qualitative_versus_numerical_identity. Accessed 20 Sept 2010
18. Wikipedia, Cultural identity (2010), Available: http://en.wikipedia.org/wiki/Cultural_identity. Accessed 23 Sept 2010
19. S. Roccas, M.B. Brewer, Social identity complexity. Pers. Soc. Psychol. Rev. **6**, 88–106 (2002)
20. J. Balmer, Corporate identity: the power and the paradox. Des. Manag. J. **6**, 39–44 (1995)
21. Poem Hunter, Quotations from Salvador Minuchin (2010), Available: http://www.poemhunter.com/quotations/famous.asp?people=Salvador%20Minuchin. Accessed 25 Oct 2010
22. C. Taylor, *Sources of the Self: The Making of the Modern Identity* (Harvard University Press, 1989)
23. G. Warnke, *After Identity: Rethinking Race, Sex, and Gender* (Cambridge University Press, Cambridge, 2007)
24. Facing History and Ourselves Foundation, *Stories of Identity: Religion, Migration, and Belonging in a Changing World* (2008)
25. R. Jenkins, *Social Identity: Key Ideas*, 3rd edn. (2008)
26. T. Miyata et al., A survey on identity management protocols and standards. Oxford Journals: IEICE Trans. Inf. Syst. (special section on new technologies and their applications of the internet III) **89**, 112–123 (2006)
27. M. Benantar, *Access Control Systems: Security, Identity Management and Trust Models* (Springer Science+Business Media, 2006)
28. A. Jøsang, S. Pope, User-centric identity management, in *Proceedings of the AusCERT Asia Pacific Information Technology Security Conference* (2005), pp. 1–6
29. P.J. Windley, *Digital Identity: Unmasking Identity Management Architecture (IMA)* (O'Reilly Media, 2005)
30. Organization for Economic Co-operation and Development (OECD), At Crossroads: Personhood and Digital Identity in the Information Society. *The Working Paper series of the OECD Directorate for Science, Technology and Industry* (2008), Available: http://www.oecd.org/LongAbstract/0,3425,en_2649_34223_40204774_119684_1_1_1,00.html. Accessed 21 May 2010
31. E. Damiani et al., Managing multiple and dependable identities. IEEE Internet Comput. (IEEE Comput Soc) **7**, 29–37 (2003)
32. Identity Gang Group—Working Group of Identity Common, Identity Gang Lexicon, Available: http://wiki.idcommons.net/Lexicon. Accessed 10 May 2010
33. Princeton University Wordnet—Lexical Database for English, Identity definition [Online], Available: http://wordnetweb.princeton.edu/perl/webwn?o2=&o0=1&o7=&o5=&o1=1&o6=&o4=&o3=&s=identity&i=1&h=0000#c. Accessed 10 May 2010
34. E. Goffman, *The Presentation of Self in Everyday Life* (1956)
35. S. Williams, This is Me digital identity and reputation on the internet, Available: http://www.slideshare.net/shirleyearley/this-is-medigital-identity-and-reputation-on-the-internet. Accessed 27 May 2010
36. Center for Democracy and Technology, Privacy principles for identity in the digital age (Draft for Comment—Version 1.4) (2007), Available: http://www.cdt.org/files/pdfs/20071201_IDPrivacyPrinciples.pdf. Accessed 28 May 2010
37. J. Hodges et al., *Glossary for the OASIS Security Assertion Markup Language (SAML) V2.0* (OASIS, 2005)
38. J. Hodges, Liberty technical glossary. *Liberty Alliance Project* (2006)
39. National Research Council—Committee on Authentication Technologies and Their Privacy Implications, Who Goes There? Authentication through the lens of privacy (2003), Available: http://books.nap.edu/catalog.php?record_id=10656#toc. Accessed 7 June 2010
40. Ofcom: Office of Communication, Social networking: a quantitative and qualitative research report into attitudes, behaviours and use (2008), Available: http://stakeholders.ofcom.org.uk/binaries/research/media-literacy/report1.pdf. Accessed 27 Aug 2010

41. D. Gollmann, Identity and location, in *Security Protocols*, ed. by B. Christianson et al., vol. 3957 (Springer, 2006), pp. 246–250
42. J. Caroll, J. Murphy, Who am I? I am Me! Identity management in a networked world, in *Proceedings of the 4th International We-B Conference* (2003)
43. International Telecommunication Union, *Digital Life*. ITU Internet Report (2006), Available: http://www.itu.int/osg/spu/publications/digitalife/docs/digital-life-web.pdf. Accessed 21 May 2010
44. Scientifc American Podcast, When the Virtual You Changes the Real You, in *Scientifc American Magazine* (2007)
45. D. Teten, S. Allen, *The Virtual Handshake: Opening Doors and Closing Deals Online* (Amacom, 2005)
46. T. Boellstorff, *Coming of Age in Second Life: An Anthropologist Explores the Virtually Human*, (Princeton University Press, 2008)
47. T. Flew, *New Media: An Introduction* (Oxford University Press, 2002)
48. N. Reichenthal, Une identité connectée. Bulletin HEC—Le Magazine des Gradués (2007), Available: http://www.gradueshec.ch/bulletins/documents/75nadine.pdf. Accessed 13 May 2010
49. S. Clauß, M. Kõhntopp, Identity management and its support for multilateral security. Comput. Netw. **37**, 205–219 (2001)
50. Organizing Committee of Digital Identity and Privacy (Human Capital and Social Innovation Technology Summit), Call for contribution to managing digital identities for education, employment and business development (2007), Available: http://events.eife-l.org/HCSIT2007/overview/dip/dip2007. Accessed 11 May 2010
51. P. Windley, Digital identity perspectives (2005), Available: http://www.windley.com/stories/2004/04/20/digitalIdentityPerspective. Accessed 5 Oct 2010
52. J. De Clercq, J. Rouault, An introduction to identity management (2004), Available: http://devresource.hp.com/drc/resources/idmgt_intro/idmgt_intro.pdf. Accessed 12 July 2007
53. R.D. Ashton et al., An organizing framework for collective identity: articulation significance of multidimensionality. Psychol. Bull. **130**, 80–114 (2004)
54. G. Ben Ayed, Consolidating fragmented identity: attributes aggregation to secure information systems. IADIS Int. J. Comput. Sci. Inf. Syst.**4**, 1–12 (2009)
55. J. Goldman, K.A. Brower, Obama's advice to aspiring politicians: be careful on facebook (2009), Available: http://www.bloomberg.com/apps/news?pid=newsarchive&sid=aL6GJ25zYajY. Accessed 4 Oct 2010
56. G. Bell, J. Gemmel, A digital life. Sci. Am. Mag. **296**, 58–65 (2007)
57. E. Hoog, Comment civiliser Internet? Le nouvel observateur (2009), pp. 20–21, Available: http://hebdo.nouvelobs.com/sommaire/les-debats/087207/comment-civiliser-internet.html. Accessed 18 May 2010
58. MyLifeBrand: social network aggregator playing catch up, Available: http://www.readwriteweb.com/archives/mylifebrand_social_network_aggregator.php. Accessed 25 May 2010
59. Mashable: the social media guide, Available: http://my.mashable.com/. Accessed 25 May 2010
60. Profilactic: a social media aggregator/life streaming service, Available: http://www.profilactic.com/. Accessed 25 May 2010
61. Snag: a social networks aggregator. Available: http://www.dapper.net/dapplications/Snag/. Accessed 25 May 2010
62. Profileomat: a shareable profile aggregator, Available: http://www.profileomat.com/. Accessed 25 May 2010
63. Naymz Features, Available: http://www.naymz.com/about.action?section=compare. Accessed 26 May 2010
64. SocialURL: showcase all your web profiles with a single, Available: http://socialurl.com/default.aspx. Accessed 26 May 2010
65. PeopleAggregator, Available: http://www.peopleaggregator.net/. Accessed 26 May 2010
66. ProfileFly: control and promote your entire online identity, Available: http://profilefly.com/. Accessed 26 May 2010

67. SocialNetwork.in, Available: http://socialnetwork.in/index.cfm. Accessed 26 May 2010
68. Mashable, Available: http://mashable.com/. Accessed 26 May 2010
69. SocialStream, Available: http://www.hcii.cmu.edu/M-HCI/2006/SocialstreamProject/index.php. Accessed 26 May 2010
70. 8hands Intelligent Social Network Aggregator, Available: http://www.logiagroup.com/socialNetworks.html. Accessed 26 May 2010
71. 8hands Software (version 0.9.135): download page, Available: http://download.cnet.com/hands/3000-12941_4-1066775.html. Accessed 26 May 2010
72. NoseRub, Available: http://noserub.com/. Accessed 26 May 2010
73. NoseRub application (Version 0.8.2): download page, Available: http://noserub.com/download/. Accessed 26 May 2010
74. Minggl, Available: http://doyou.minggl.com/. Accessed 26 May 2010
75. Minggl Download Page (Firefox add-on), Available: http://doyou.minggl.com/welcome/install/. Accessed 26 May 2010
76. G. Ben Ayed, S. Ghernaouti-Hélie, Digital identity attributes cohesion: major issues and challenges for e-services access control, in *Presented at the International Conference on Information Technology and E-service (ICITeS'2011)*, Sousse, Tunisia, 2011
77. Wink: people search engine, Available: http://wink.com/. Accessed 25 May 2010
78. Secondbrain: save, share and discover great bookmarks, Available: http://secondbrain.com/. Accessed 26 May 2010
79. Wikipadia, Context awareness (2011), Available: http://en.wikipedia.org/wiki/Context_awareness. Accessed 23 Aug 2011
80. A.R. Galloway, E. Thacker, *The Exploit—A Theory of Networks* (University of Minnesota Press, 2008)
81. A. Ross , A.K. Jain, Biometrics: when identity matters, in *Advances in Biometric Person Authentication,* vol. 3338/2005 (Springer, 2005), pp. 137–149
82. J. Fildes, Taking control of your digital id (2006), Available: http://news.bbc.co.uk/2/hi/technology/6102694.stm. Accessed 22 May 2010
83. M. Fischetti, Scoring your identity: new tactics root out the false use of personal data. Sci. Am. 27–28 (2007)
84. P. Cochrane, Forward of the book, in *Digital Identity Management: Perspectives on the Technological, Business and Social Implications,* ed. by D.G.W. Birch (Gower Publishing Limited, 2007)
85. A. Henderson, Practical action: federation and mobility, in *Digital Identity Management: Perspectives on the Technological, Business and Social Implications,* ed. by D.G.W. Birch (Gower Publishing Limited, 2007), pp. 72–88
86. Facebook, Facebook statistics (2011), Available: http://www.facebook.com/press/info.php?statistics. Accessed 13 Nov 2011
87. J. Goodchild, The robin sage experiment: fake profile fools security pros (2010), Available: http://www.networkworld.com/news/2010/070810-the-robin-sage-experiment-fake.html?t51hb. Accessed 8 Nov 2011
88. B. Ferran, Une fausse chercheuse dupe des experts en sécurité (2010), Available: http://www.lefigaro.fr/web/2010/07/24/01022-20100724ARTFIG00481-une-fausse-chercheuse-dupe-des-experts-en-securite.php. Accessed 8 Nov 2011
89. G. Tissier et al., Les marchés noirs de la cybercriminalité. Compagnie Européenne d'Intelligence Stratégique (CEIS) (2011)
90. Wikipedia, E Pluribus Unum (2011), Available: http://en.wikipedia.org/wiki/Out_of_Many_One. Accessed 25 Oct 2010
91. Wikipedia, Great seal of the United States (2010), Available: http://en.wikipedia.org/wiki/Great_Seal_of_the_United_States. Accessed 25 Oct 2010
92. S.L. Garfinkel, Information of the World. UNITE! Sci. Am. Mag. 82–87 (2008)
93. K. Cukier, *All Too Much: Monstrous Amounts of Data* (The Economist, 2010), Available: http://www.economist.com/specialreports/displaystory.cfm?story_id=15557421. Accessed 13 May 2010

94. K. Cukier, *The Data Deluge: Businesses, Governments and Society are Only Starting to Tap its Vast Potential* (The Economist, 2010). Available: http://www.economist.com/opinion/displaystory.cfm?story_id=15579717. Accessed 13 May 2010
95. Garlik, Garlik's DataPatrol (2010), Available: http://www.garlik.com. Accessed 6 Aug 2010
96. Mypublicinfo, Mypublicinfo: identity protection services (2010), Available: http://www.mypublicinfo.com. Accessed 26 Aug 2010
97. G. Ben Ayed, S. Ghernaouti-Hélie, Digital identity attributes cohesion to access e-services: major issues and challenges in digital society. J. E-Technol. **2** (2011)
98. G. Ben Ayed, S. Ghernaouti-Hélie, Claim-based digital identity fusion to access e-services: major issues and challenges in digital society, in *International Conference on Information and Computer Applications (ICICA 2011)*, Dubai, UAE, 2011
99. J. Harris, S. Kamvar, We feel fine project: an exploration of human emotions, in six movements, Available: http://wefeelfine.org/. Accessed 15 May 2010
100. J. Harris, S. Kamvar, we feel fine project: blogs data [Online], Available: http://www.wefeelfine.org/data/files/feelings.txt. Accessed 25 May 2010
101. M. Hansen, B. Rubin, Listening Post Project, Available: http://www.earstudio.com/projects/listeningpost.htm. Accessed 25 May 2010
102. S. Billard, Référencement, Design et Cie (2008), Available: http://s.billard.free.fr/referencement/?2008/10/27/521-123people-moteur-de-recherche-de-personnes. Accessed 25 Oct 2010
103. Google, About News Archive Search, Available: http://news.google.com/archivesearch/about.html. Accessed 19 May 2010
104. H. McCallum-Bayliss, Identity resolution in a global environment: fishing for people in a sea of names. IEEE IT Prof. **6**, 21–26 (2004)
105. K. Cukier, Cupo (The Economist, 2010), Available: http://www.economist.com/specialreports/displaystory.cfm?story_id=15557507. Accessed 13 May 2010
106. R. Bowes, A Quick Hack to Download Facebook URLs from Facebook Directory (2010), Available: http://www.skullsecurity.org/blogdata/facebook.rb. Accessed 29 Aug 2010
107. R. Callow, 100 million Facebook users added to a publicly available torrent file (2010), Available: Sync-Blog http://www.sync-blog.com/sync/2010/07/100-million-facebook-users-added-to-a-publicly-available-torrent-file.html. Accessed 29 Aug 2010
108. R. Bowes, Return of the Facebook snatchers (2010), Available: SkullSecurity Blog http://www.skullsecurity.org/blog/?p=887. Accessed 29 Aug 2010
109. R. Paul, Leaked. Data of 100 M Facebook Users Came from Public Info (2010), Available: http://arstechnica.com/security/news/2010/07/leaked-data-of-100m-facebook-users-came-from-public-info.ars. Accessed 14 Oct 2010
110. Wikipedia, Digital native, Available: http://en.wikipedia.org/wiki/Digital_native. Accessed 18 May 2010
111. J. Palfrey, U. Gasser, *Born Digital: Understanding the First Generation of Digital Natives* (Basic Books, 2008)
112. J. Howe, Crowdsourcing: Why the Power of the Crowd Is Driving the Future of Business, Crown Business, (2008)
113. United Nations Population Division—Department of Economic and Social Affairs (1999) The World at Six Billion. Available: http://www.un.org/esa/population/publications/sixbillion/sixbillion.htm Accessed: May 18 2010
114. D. Jefferies, Preparing for the Digital Afterlife, in The Guardian Newspaper, ed, 2009
115. K. Cukier, Needle in a Haystack: The uses of Information about Information. The Economist (February 23–March 5). Available: http://www.economist.com/specialreports/displaystory.cfm?story_id=15557497 Accessed: May 13, 2010
116. K. Warwick, *Cyborg identity in digital identity management: perspectives on the technological, business and social implications,* ed by D. G. W. Birch (Gower Publishing Limited 2007), pp. 227–238
117. K. Cukier, Data, data everywhere. The economist (February 23rd–March 5th). Available: http://www.economist.com/specialreports/displaystory.cfm?story_id=15557443 Accessed: May 13, 2010

118. D. Durand, Nouvelles vidéos sur Youtube: +50'000 heures chaque jour ! évolution depuis 2007. Available: http://www.zdnet.fr/blogs/media-tech/nouvelles-videos-sur-youtube-50-000-heures-chaque-jour-evolution-depuis-2007-39756042.htm Accessed: Nov. 12, 2010
119. K. Cukier, The open society: governments are letting in the light. The economist (February 23–March 5). Available: http://www.economist.com/specialreports/displaystory.cfm?story_id=15557477 Accessed: May 13, 2010
120. K. Cukier, (2010) Clicking for gold: how internet companies profit from data on the web. The economist (February 23–March 5). Available: http://www.economist.com/specialreports/displaystory.cfm?story_id=15557431 Accessed: May 13, 2010
121. K. Cukier, (2010) New rules for big data: regulators are having to rethink their brief. The economist (February 23–March 5). Available: http://www.economist.com/specialreports/displaystory.cfm?story_id=15557487 Accessed: May 13, 2010
122. Spokeo: people finder, Available: http://www.spokeo.com/. Accessed 26 May 2010
123. J.J. Luna, *How to Be Invisible: A Step-by-Step Guide to Protecting Your Assets, Your Identity, and Your Life* (Farrar, Straus, and Giroux, New York, 2000)

Chapter 3
Digital Identity Management

There's the common basis for communication.
A new language. An intersystem language.
But a language only those machines can understand.
(Colossus: The Forbin Project, 1970)

In this chapter, we present a literature review on the definitions of Digital Identity Management (DigIdM), various origins of identity silos, and digital identity management technical models. We provide also a comparison between digital identity management technical models and then we explain the basics of a new technical approach that is based on global Web digital identity management. We highlight the contributions of user-centric digital identity management. Finally, we explain a metadata-based approach to make digital identity less visible in order to give users more control on persistent digital identity.

3.1 Digital Identity Management: Basics

The move to the virtual world brings with it new security risks. Increasing number of regulations in US and EU is driving the need to ensure security. The author [1] adds that security models should lay on identity management and identity must become persistent through any given process spanning multiple applications and organizations. He stresses that identity is a predicate for corporate governance, security, regulatory compliance, risks and liability management, and other core business functions. Digital identity management will emerge as a pervasive infrastructure, within, between, and across organizational boundaries [1]. Besides considering digital identity management as one of the security pillar [2], it is also considered as one of the major enablers of e-business [3]. It should provide information security, privacy and trust in order to allow further network boundaries expansion, access points, innovative practices and technologies [3]. However, actually Digital identity management is still suffering from multiple complexities [2, 4].

G. Ben Ayed, *Architecting User-Centric Privacy-as-a-Set-of-Services*, Springer Theses, DOI: 10.1007/978-3-319-08231-8_3,

DigIdM is considered as a critical security component. The author [5] stresses potential results of a bad identity management: (1) fragmented point solutions; (2) failure to deliver real business value; (3) failure to leverage existing investments and infrastructure; (4) dilution of identity management initiatives over time; and (5) increasingly difficult funding for further initiatives [5]. Issues and elements of DigIdM are classified into a set of 'components' [6] or 'stacks' [1]. Components are classified into three sides: (1) technology side includes identity tools as an interface between standards and systems; (2) business side includes identity business as an interface between systems and rules; and (3) society side includes identity management scheme as an interface between rules and stuff [6].

3.2 Taxonomy of Digital Identity Management Definitions

Various DigIdM definitions are suggested in the literature because they are defined and seen from different perspectives. We provide and discuss major definitions of DigIdM in literature review and we highlight that intra information system DigIdM has vertical silos-focus and that of networked information systems requires horizontal end-to-end and processes-focus. In addition, most of DigIdM definitions take into consideration a composition of more than one perspective, which makes their dissociation very challenging task. We use the term definition-focus to explain that DigIdM is defined on the basis of a prime focus perspective, which we consider it in order to establish taxonomy of DigIdM definitions. Below, we classify DigIdM definitions on the basis of three definition-focus perspectives as follows.

3.2.1 DigIdM Security System and Technical Definition-Focus

The author [1] defines DigIdM as a set of access control system's technical requirements. He says that identity management is "to encompass not only requirements to correctly identify who a person is, but also the manifestations of that knowledge through SSO, account provisioning, authentication and authorization". The author [7] defines DigIdM as a set of technical models, which are classified into four categories based on identity's scope. He argues that identity management paradigms in computing are analogous to real-life practices. In fact, the scope of an individual identity varies from one person to another. A person may be known only to his or her family, immediate neighbors, or a workplace; another person can be known throughout his or her locality or a much bigger geography; while another person is known over the globe. The scope of identity in computing follows the same logic: (1) local identity model such as local registry management of users; (2) network identity model such as cross-domain Kerberos and PKI cross-certification implementations; (3) federated identity in which cross organizational trust or circle of trust is a foundation; and (4) global Web identity such as meta-directory, virtual-directory, and OASIS Extensible Resource Identifier (XRI) and Extensible Data Interchange (XDI) infrastructure implementations [7]. However, there is a distinction between identity federation and federated identity. The first is a conceptual model and the second is an implementation of that model. Moreover, meta-directory and virtual-directory are respectively implementations of meta-centralization and virtual-centralization conceptual models [8]. Authors of ITU report points out that requirements from subject perspective is different from that of organizational perspective and therefore DigIdM system should defined separately from two perspectives. Moreover, the same authors provide DigIdM authentication-purpose definition in which claim-based administration, verification, authentication, and revocation should be properly supported by a number of different technologies such as electronic signatures, password synchronization, PKI, federated identity systems, interoperability standards, and directories [9]. With access control-based management definition, the author [10] states that Identity and Access Management (I&AM) systems allows organizations to manage employees' and customers' digital identity attributes and access rights to central enterprise directory. The same author adds that in order to respond to networked information systems requirements, I&AM systems have developed into federated identity management (FIM) systems, which lay on FIM standards, such as OASIS SAML, Liberty Alliance, and WS-Federation; cross-domain SSO; and circle-of-trust relationship. While, the author [11] defines 'identity management architecture' as a framework of identity management solution that has several key components: enterprise information architecture, permission and policy management, enterprise directory services, user authentication, user provisioning, and workflow. However, DigIdM is considered as a tool for automating manual user administration processes [12].

3.2.2 DigIdM Security Management Definition-Focus

The author [1] focuses on managerial aspect in defining DigIdM. He considers DigIdM as "the process of creating, managing, using, and eventually destroying records that identify a person, a car, a computer, a piece of land, etc." A broader definition is also suggested as the need to identify subjects while considering multiple associations and roles and the management of subject's information over time and across the enterprise" [13]. The author [1] defines identity management as an architecture of interrelated five blocks: process architecture, data architecture, technical reference architecture, policies and interoperability framework. He stresses that management, policy, and political issues are things that stand in the way of identity management success. He does not undervalue technical issues but risks are lower when digital identity management related technologies such as cryptography, authentication, authorization, identity provisioning, directories, digital rights management, identity federation, and interoperability standards fit into an overall identity management strategy [1]. ITU Focus Group on Identity Management provides a definition of identity management as "the management by trusted providers of trusted attributes of a subject" but they don't clearly explain concepts of trusted providers and attributes. However, the group has identified DigIdM system's critical requirements from different perspectives: technical mechanism and protocols, best practice or guidelines specification, performance specification, business models, assumptions (e.g. scalability), administrative mechanism, and national mandate [14]. The same group initiates identity assurance as way to manage risks associated with DigIdM [14]. Authors [15] provides a broad definition of identity management as "definitions and lifecycle management for digital identities and profiles, as well as environments for exchanging and validating such information". Whereas, authors [16] stress how the scope of identity management has evolved from that of intra-information system level into that of inter-information systems level. Traditionally DigIdM has been concerned with "managing an organization's employees to ensure that their authentication and authorization information is consistent and synchronized within organization's information system". Currently, DigIdM is the "ability to federate identity across organizations while maintaining clear trust, liability, and cost responsibilities" [16]. In addition, the author [17] highlights that DigIdM in an open interconnected information systems lies on access controls risk management. He considers "controls, where what you can do is based on who you are, are fundamental to managing risk". Risks could be financial, information security, and/or compliance with legal and regulatory requirements. In parallel, DigIdM is defined as "centralized policy-based management of all information required for access to enterprise systems by people, host, programs, or other resources" but the definition is limited to intra-information system [18]. However, DigIdM should be perceived from strategic point of view, therefore it is more than "solving technical issues or dealing with compliance requirements- rather than from a strategic point of view, which is business-driven and outcome-based" [5]. From the point of view of identity attributes rules-based control, OECD [19] describes DigIdM is to be constructed on four interdependent levels: knowledge of identity nature or 'properties of identity' on which lays data protection, which in its turn guarantees accountability. Ultimately accountability is the pillar of trust management. The OECD report details only the two first levels, which drove the definition into attributes control domain, specifically identifiers and claims management. Identifiers could be a subject name that is comprehensible by a human or a machine, such as person's first-name; and a claim is a statement about the subject's behavior or possessions, such as the subject holds a CC Bank credit card (# 123456790). In order to have subject's control over identifiers and claims, both legal and technical mechanisms are required. If the mechanisms are not successfully addressed, many issues the identity will face, such as identity fraud and privacy [19]. Wikipedia perceives also DigIdM from access control point of view and specifically from the administrative as "a wide administrative area that identifies users in a system and defines restrictions on established identities" but it does not clearly explain details and/or steps behind administrative area. A vision is needed to have DigIdM evolve [6]. A well-defined identity management strategy can improve the agility of IT infrastructure, allowing organizations to be more responsive and resilient to the rapid pace of change [12]. The identity management infrastructure has improved the ability to respond quickly to changes throughout the organization [12].

3.2.3 DigIdM User-Supremacy Definition-Focus

The concept of user-centricity arises out of giving subject convenience and sovereignty over personal data. Technically, in an identity federation setting, a user-centric DigIdM system incorporates three components: (1) identity provider which stores identity attributes and authenticates the subject; (2) service provider, called also relaying party; and (3) identity selector that allows subjects to choose which identity provider to use and what information to disclose to a particular service provider [19]. From privacy-preserving perspective, 'attributes management' system are to be developed on how "to ensure that no part of a system can aggregate an individual's private attributes" [20]. Moreover, user-controlled DigIdM by which subjects can choose the appropriate partial identity according to the current application requirement. The subject could also manage the plurality of accounts and passwords and allows keeping track of which digital identity attributes that have been disclosed to and processed by whom. Pseudonyms could be used to prevent other parties' undesired context-spanning linkage and profiling [21]. Users should have a stronger position against service providers and for this reason DigIdM has changed into digital identity assurance [22].

3.3 From Vertical into Horizontal Management

Yesterday, companies defined their organizational and operational models based on functional areas of the business, which inherently focuses on vertical silos. Vertical silos lead to ad-hoc processes that are fragmented rather than integrated and hidden processes that are difficult to see, manage, or predict. In addition, information technology might be aligned to the vertically siloed departments for development of systems, thereby reinforcing complexities [23]. However, today, horizontal DigIdM processes are needed in the open Web of organizations setting. We define DigIdM processes are a series of repeated steps and actions to create tangible value for users. The author [23] quantifies the value and points out that the value stream includes end-to-end horizontal processes that cut across functional organizational boundaries with interaction from the users and various value-chain partners. The value of a process-oriented approach includes: (1) alignment to business strategy: business processes capture the essence of the business strategy with respect to process priorities. A process-oriented approach helps focus the operational alignment and execution to the overall business strategy; and (2) user-focus: users are the key drivers and primary beneficiaries of effective business process management. Extra value inevitably results when business processes align to the user's needs and provides a unifying view of customer information [23].

Many authors provide process-oriented DigIdM definitions and highlight the current need of DigIdM processes to secure and protect digital identity attributes. The author [17] considers DigIdM as a set of processes and technologies involved in implementing access controls. In parallel, ITU Focus Group defines DigIdM as 'the process of secure management of identity information (e.g., credentials, identifiers, attributes, and reputations). Narrowly, the group points out that identity management is the technology behind establishing, modifying, suspending, archiving or terminating identity information; recognizing partial identities that represent entities in a specific context or role; establishing and assessing trust between entities; and the discovery (location) of an entity's identity information (e.g., authoritative IdP that is legally responsible for maintaining identifiers, credentials and some or all of the entity's attributes) [4]. DigIdM "streamlines various business process that deal with managing all forms of identities in an organization, from enrollment to retirement" [5]. The definition ignores and contradicts specific indication of involving processes, however, authors of ITU report sub-divide DigIdM process into three sub-processes: digital identity verification, subject's authentication, and digital identity revocation [9].

3.4 Digital Identity Management Technical Models

There are many identity management emerging standards in the field of DigIdM. The liberty alliance specifications define the protocol messages, profiles, and processing rules for identity federation and management. The SAML provides a set of XML and SOAP-based services, protocols, and formats for exchanging authentication and authorization information. Other evolving standards and ongoing projects in the field of digital identity management are: SXIP [24]; LID [25]; XRI/XDI [26]; OpenID [27]; YADIS [28]; and Windows CardSpace [29]. Several standardization bodies and similar organizations are working on identity management standardization such as ITU-T [30]; well-established open communities such as IETF [31], Kantara Initiative [32] (formerly Liberty Alliance Project [33]), Shibboleth Project [34], Bandit Project [35], Higgins Project [36]; and regional projects such as European Daidalos Framework Project [36] and European PRIME Project [37].

Identity Management inherently involves sets of information exchanges between two parties according to some protocol known between them. It is a standard information exchange model where a requesting or asserting party conveys an assertion or query message to another party as the basis for some response or action that involves identity. In most but not all cases, there will be some kind of response message or action. A person wearing a nametag in a public space is

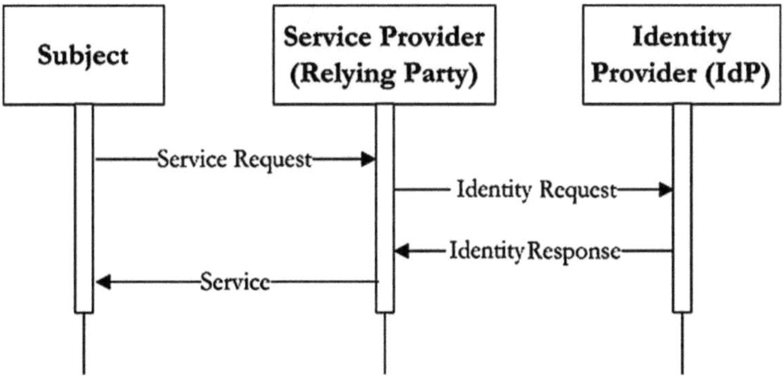

Fig. 3.1 Common structured identity management model [38]

an example of an identity assertion where there may be no response message or action [38]. The parties may be any kind of entity—real persons, organizations or institutions, or any of a myriad kind of physical or virtual objects in the form of peripheral terminal devices and sensors, network equipment, actively tagged physical objects (e.g., using RFIDs or optical codes), passively tagged objects, geospatial constructs, software, or multimedia content of all kinds. Depending on the level of assurance desired, that party makes a decision to engage in sets of additional query-response messages with an identity provider (which may be the relying party itself or another party within a federation or alliance relationship) to validate the assertion via credentials, identifier, attribute, and pattern identity services. The result is a simple, near universal Identity Management model depicted in Fig. 3.1.

The author [7] stresses the importance of identity scope and provides a taxonomy of identity models based on the scope of an identity. He adds that identity management paradigms in computing are analogous to real-life practices. In fact, in real-life the scope of an individual identity varies from one person to another. An individual may be known only to his or her family, immediate neighbors, or a workplace; another person can be known throughout his or her locality or a much bigger geography; while another person is known over the globe. The scope of identity in computing follows the same logic. The same author discusses the local identity scope such as meta-directory, followed by network scope, and then the global scope such as identity federation. We classify digital identity management technical models into two classes on the basis of identity scope [7] of digital identity management. The two classes are centralization, which is subdivided into two sub-categories, and federation. We borrow IdP, SP, and AA identity federation-specific concepts to explain the meta-centralization and virtual-centralization models for the following main two reasons: (1) to better and clearly explain and compare between the technical models with the same parlance; and (2) to highlight communication and attributes convey between providers. We provide a description of technical models by focusing on data exchange between the stakeholders and describe issues related to each of them.

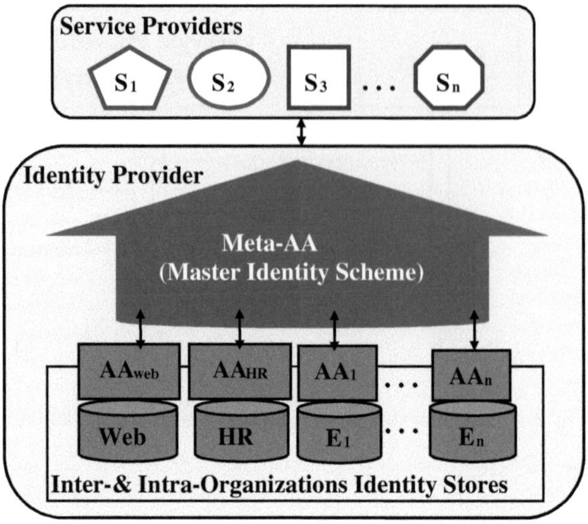

Fig. 3.2 A high-level description of the meta-centralization model

3.4.1 DigIdM Centralization: Meta-directory Technical Model

The meta-directory defines a centralized repository that is built directly on the top of the existing systems. It also provides a unique consolidated and centralized view by unifying distributed attributes across different identity stores. In Fig. 3.2, Meta-AA represents authority that manages the meta-directory and plays the role of a middleware between SPs and AAs. Within services provider envelope, we represent different types of services by different shapes with colored borders. AA represents authority that manages the repository and provides the requested attributes to Meta-AA. However, Meta-AA manages a unique master account for all participating AAs. In this structure, a user is in one-to-many relationship with his sets of attributes in the underlying AAs. IdP manages all the identity attributes provided by AAs and Meta-AA and conveys attributes to SPs through namespace connector. The authors [1, 7] point out that Meta-AA administers two main services: attributes aggregation (push up) and attributes synchronization (push down). In one hand, identity attributes aggregation process allows collecting all the attributes from different AAs and pushing them up to the central Meta-AA. Technically, a join operation is performed to copy attributes from various underlying directories that are keyed by joint points through a join-link. These links are configured separately to filter the desired attributes. In the other hand, identity attributes synchronization propagates and pushes down the changes from Meta-AA to AAs. Meta-AA maintains a master identity scheme, which comprises either all the attributes provided by AAs or only some of the attributes that were considered relevant during system configuration.

The author [7] suggests two ways to specify and implement the 'master identity scheme': a unified identity-representation scheme and a decoupled identity-representation scheme. In the unified scheme, master identity scheme, which is maintained by Meta-AA, encapsulates a superset of all identity attributes. Each AA may introduce attributes and contribute to master identity but AA is aware of only a subset of the common identity attributes. Multi-valued attributes on master identity scheme is allowed because the same attribute might have different values within different identity stores. Note that, attributes with no values that are assigned to them may be permitted within master identity scheme. However, a mapping may be needed to relate an attribute defined on Meta-AA to the corresponding attributes maintained by AAs. AA might have to manage new defined attributes, which might be not visible to Meta-AA and not common to other AAs, hence, a dynamic redefinition of the schema and a full reconfiguration of the meta-directory system are needed. Here, Meta-AA maintains all attributes in a unique identity vault and attributes are replicated piecewise across identity stores. Attribute retrieval operations, therefore, can be send to Meta-AA and do not require involving AAs. In the decoupled scheme, only a fixed set of attributes are maintained by Meta-AA and AA-specific attributes are not visible to Meta-AA. Adding new identity store would not impact the master identity scheme. Here, the scheme requires only one setup at the meta-directory but in the unified scheme, it requires one at the meta-directory and another at identity stores. Data updates policies are also to be taken into consideration; If changes are allowed at Meta-AA and AAs levels, synchronization becomes complex. If the changes are allowed only at the Meta-AA level, complex authorization policy can ensure that only identity owners can modify accounts information [1].

3.4.2 DigIdM Centralization: Virtual-Directory Technical Model

Virtual-directory participates in tightly coupled structure to create and enable a single integrated logical view of attributes within multiple directories [1, 7]. Virtual-AA is a querying authority that manages virtual-directory and performs real-time attributes pooling from disparate trusted AAs named authoritative sources as shown in Fig. 3.3.

Moreover, Virtual-AA is represented in a discontinued line box to highlight the fact that virtual-directory is a logic and non-physical directory that disappears instantly when the query is completed. Attributes mapping is processed while all the identity attributes are kept intact in the underlying repositories. The main difference between Virtual-AA mapping approach and that enabled by Meta-AA is that Virtual-AA is not keeping data in a central attributes repository. A query to the virtual-directory is turned by Virtual-AA into multiple queries distributed over the participating AAs. Virtual-AA receives queries and directs them to the appropriate AAs and then the result is sent by IdP to SPs through application programming interface (API). Virtual-AA retrieves and updates attributes maintained by multiple AAs simultaneously through an initial setup of a collect operation. Virtual-AA

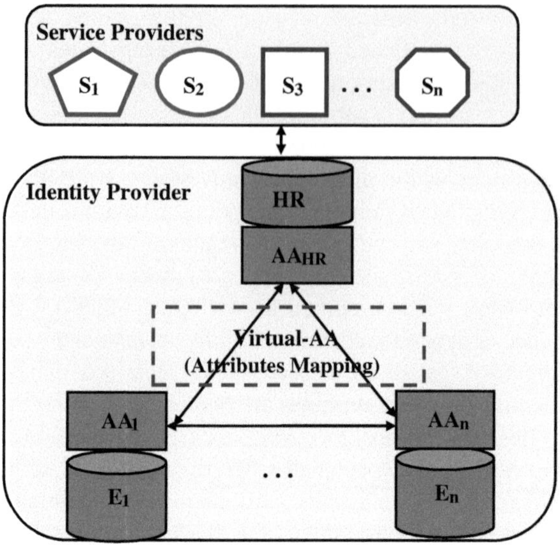

Fig. 3.3 A high-level description of the virtual-centralization model

uses one attribute as the join-key in order to match entries across different identity directories. The join-key is the name of an attribute that is used as the common link between identity stores. Mapping identity attributes across all AAs, however, creates management complexities associated with n-wise mapping issue [7]. Moreover, attributes updates may require synchronization across multiple directories. It is helpful to consider automated synchronization; otherwise, complexities and data errors are very likely to increase. The author [1] recommends virtual-directory use in cases where real-time access to frequently changing attributes is important.

3.4.3 DigIdM Federation Technical Model

Microsoft Passport is a cross-domain web single sign on that allows users to manage their digital identity attributes, pseudonyms, and credentials by themselves. It is considered as an implementation of centralized technical model. Rolph Oppliger with eSECURITY Technologies, Switzerland, provides in a paper [39] an overview of Web-based single sign-in (SSI) service .NET Passport and addresses the question whether the service meets the users' identity management requirement on the Web. In his paper [40], the author examines the feasibility and desirability of using the Microsoft Passport service for client authentication and authorization. He concludes that the adoption of Microsoft passport challenged by the lack of trust, control, and privacy; and the proliferation of other identity management models. Users are usually faced with dilemma of balancing security and convenience when creating multiple accounts. Should they maintain a list of usernames

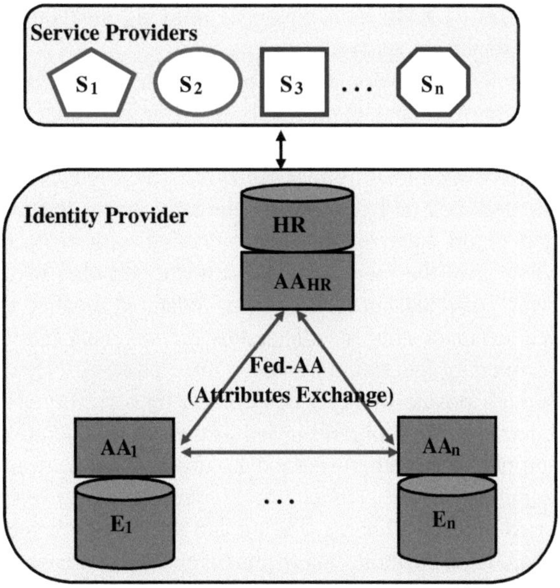

Fig. 3.4 A high-level description of the identity federation model

and passwords or use the same username and password for every account? On the enterprise side, the need of users' data consolidation gave the birth to federated identity management (FIM). FIM aims to allow organizations to securely share confidential user identities with trusted ones without requiring the users to re-enter their usernames and passwords.

Coordinating and integrating business processes with partners is a complex dilemma faced by most large enterprises. Identity federation addresses this cross domain security challenge and allows tying distributed applications together. The term federation is used to imply collaboration between loosely coupled sovereign organizations [41]. In addition, identity federation holds the promise of delivering significant benefits to both users and organizations [41]. Organizations involved in identity federations establish trusted relationships with other parties to allow users and systems accessing resources available across information systems. Based on glossaries of [42, 43], 'federated identity' defines an agreement between the providers on a set of attributes to refer to the user. While, 'identity federation' is the act of creating federated identity on behalf of the user. The authors [1, 7] mention that federated identity enables controlled linkages of attributes between heterogeneous systems while attributes stay locally. Fed-AA is the software, manager, and authority that administers the exchange of AAs' attributes in a form of assertions between IdP and SPs. OpenID uses multiple IdPs [44]. The exchange of assertions is represented in Fig. 3.4 by the blue-colored arrows. The same authors stressed that establishing and maintaining trust across organizations is a core of identity federation. Specifically, identity federation can only communicate trust between organizations but it cannot establish it. As a consequence, attributes may ultimately be required to adhere to a common representation scheme and semantics.

The use of XML as a means of defining attributes can ease interoperability and acceptance across organizations.

In the discussion transcript [45], the author distinguishes between federated identity-management protocols and federated identity management. Federated identity-management protocols refer to digital identity systems such as Liberty Alliance and federated identity management is the digital identity system that the company has designed and developed internally. From the merchant perspective, a discussion group member points that dealing with protocols, such as Liberty Alliance, would give the merchant access to the growing network of customer profile data being collected but dealing with federated identity management, customer profile data would be simply collected by the merchant himself [45]. The author [1] classifies identity federation based on three patterns: (1) ad-hoc federation is established through private bilateral agreements between organizations; in (2) hub-and-spoke federation, large organizations form private federation islands; (3) identity federation network is characterized by the formation of an independent member-owned identity platform. The author [7] presents three federation topologies categorized based on local user registration and attributes schemes: (1) local profiling topology where local attributes management and user's registration are at home organizations and other organizations would be aware of such registration only when attributes are exchanged across them; (2) the distributed profiling topology: an organization may acquire, through additional registration, new attributes from specific organizations. Thus, identity attributes may be duplicated; (3) third party profiling scheme: a designated third party within the established federation is tasked to manage the attributes. The third party knows attributes that are common to all or to a subset of the organizations and those that are relevant to specific ones. Organizations have to establish and manage trust with only the third party, who would take care of attribute synchronization. In addition, [46] proposes in the identity federation context three association methods that could be used for aggregation: (1) contextual association method allows multiple SAML assertions to be simultaneously propagated to providers by the same user. The attributes on assertions will be linked by a context; (2) identifier sharing method permits user identifier that is used at IdP1 to be transmitted to IdP2 through user's authentication request. If IdP2 re-authenticates the user via an identifier already knows by IdP2, the IdP1 would know that both identifiers are valid for the same user. Here IdP2 maintains user attributes. If the user is not registered at IdP2, which may need to store user attributes, it could use the identifier sent by IdP1 as an identifier in the creation of the user account locally without re-authenticating the entity; (3) identity federation method allows IdP to create a new identifier for identity that is maintained anonymously with pseudonym. Accounts may be aggregated by passing the identifier from one IdP to another by applying identifier sharing method. The author presents in his paper [41] two typical modes of federation: browser-based and document-based. Browser-based federation enable authenticated user, through SSO, to move from one web security domain to another without needing to provide credentials again. By contrast, document-based federation is based on the use of XML documents transported between two security domains leveraging Web service standards.

An assertion may also contain an expression of preferred validation or a "delegation." Delegations are very important as a meant to accommodate situations where the identity is controlled within a consensual sharing relationship such as co-ownership among spouses, by an organization or institution because of an employment or other formal relationship, where a person may have diminished capacity or be a minor, where a decision to delegate authority occurs, or where objects are involved. An assertion may also be one of anonymity or pseudonymity. In such cases, the level of identity assurance is dependent on other extrinsic factors that the Relying Party would need to undertake such as examining attributes of the communication or pattern analysis. Anonymity and pseudonymity are frequently manifested where the kind of activity involved is so trivial that any kind of identity management overhead is not needed or desired [38]. 'Identity system' represents any program or framework that involves the collection, authentication, or use of identity or linked information. 'Linked information' are other facts about an individual, such as transactional, shopping or travel behavior, tied to an identity. 'Account Linkage' is a method of relating accounts at two different providers that represent the same entity so that the providers can communicate about the entity. Account linkage can be established through the sharing of attributes or through identity federation. The identity of an entity is said to be 'federated' between a set of providers when there is an association between a set of identifiers and attributes of that entity. 'Identity federation' is the act of creating a federated identity on behalf of the entity. 'Circle of Trust' (CoT) is a federation of service providers and identity providers that have business relationships. 'Policy Decision Point' is a system entity that evaluates decision requests in light of applicable policy. 'Policy Enforcement Point' is a system entity that performs access control by making decision requests and enforcing authorization decisions.

3.4.4 Comparing DigIdM Technical Models

We present the result of comparison between meta-centralization, virtual-centralization, and identity federation based on ten factors as shown in Table 3.1. Meta-centralization is a two-level model since it requires an additional physical store that plays the role of an identity vault. Ideally, the identity manager would have only one access point, instead of multi-directories access points, to maintain identity attributes, quickly locate, and eliminate attributes duplications. The identity vault would enforce an element of control within an organization under a single authority and unifies attributes management processes [7, 47]. Moreover, the vault is considered as single point of reference; whether we change directory vendors, modify system implementations, or reorganize attributes, SP still query a single source [1]. Meta-centralization is considered with a low risk of store unreliability and data unavailability since attributes have been replicated. In the other hand, having the vault would increase risks of denial service attack and attributes exposure. While Figs. 3.2, 3.3 and 3.4 show different types of attributes authorities and two

Table 3.1 Aggregation models comparison

Factors	Meta-centralization	Virtual-centralization	Identity federation
Storage-based levels	Two levels: meta-directory and identity stores	One level: identity stores	One level: identity stores
Admin. and access points	Single	Multiple	Multiple
Risk of stores unreliability	Low	High	High
Risk of denial service attack and attributes exposure	High	Low	Low
View creation of identity infrastructure	Single	Single	No
Attributes authorities	Meta-AA and AAs	Virtual-AA and AAs	Fed-AA and AAs
Supported IdPs	Single	Single	Single /Multiple
System critical pre-requisite	Attributes duplication, synchronization and master identity scheme setup	Authoritative sources availability	Trust communication
Attributes governance/ ownership issues	High	Low	Low
Global scalability	No	No	Yes

providers, [46] mentions identifier usage by multiple IdPs in identity federation. Each implementation and configuration of the three models has critical pre-requisites, as shown in Table 3.1. The meta-directory requires attributes replication from all the underlying identity stores and synchronization capabilities. The author [7] explains that unified or decoupled attributes schemes should be selected before configuring the meta-directory and places emphasis on configuration complexities of attributes updates policies. Moreover, in unified scheme, attributes ownership and governance could be a very complex issue.

The landscape in virtual-centralization and identity federation shows multiple administration access points and attributes distributed across multiple identity stores. The landscape would inevitably lower attributes exposure risk and governance issues but increase identity stores unavailability risks. While, virtual-centralization requires a high availability of trusted attributes stores, identity federation needs trust communication between stores. While, meta-directory and virtual-directory create a single view of identity infrastructure, identity federation does not; rather, identity stores cooperatively solve identity tasks. Virtual-directory has a better scalability property over meta-directory because it does not centrally storing identity attributes but only federated identity has the most potential of global scalability [1, 47]. The author [1] adds that meta-centralization and third party profiling topology of identity federation cannot scale to the extent to which they can accommodate a large number of worldwide identity stores. Virtual-centralization and identity federation do not violate internal or external regulations governing identity attributes because identity attributes stay at home identity stores. Within, identity federation, local profiling topology is well suited when identity attributes are well defined and understood by other organizations; otherwise it would not offer global scalability. Distributed profiling topology [7] may offer global scalability but attributes duplication may pose synchronization issue. The topology offers some flexibility in term of attributes ownership since there is a separation of concerns when managing attributes among organizations. In the third party profiling topology [7], scalability issue can be a serious concern when a very large population of organizations may contend over the single third party to retrieve and update all identity attributes. Given the intense focus on privacy and personal control of digital identities, and the high value of customer information that is often housed within the existing identity infrastructure, organization could not collaborate on creating and maintaining a universal, shared point of identity information [41]. Using a single a centralized identity solution for multiple purposes creates a single target for privacy and security abuses by identity thieves, terrorists, government, business, and others. The benefits of collecting and using identity, authentication, and linked information should be weighed against the risks to privacy [44].

3.4.5 XRI and Social Web Technical Approach

The following terms are taken from OASIS publications [48–50]. "Identifier" is anything that is being identified from all other things within its scope of identity. "Data" is considered as any information that when associated with an identifier becomes a "resource". Seven types of data have been specified: authentication data, control data, link data, query data, registration data, resolution data, and trust data. However, "data authority" (DA) is a resource that asserts authority over data and its association with one or more identifiers. DA can delegate control over data to another DA, who becomes a "delegated data authority" (DDA). People and organizations are types of data authorities, who can delegate authority to software agents and applications. "Identifier authority" (IA) is a type of data authority that assigns identifiers, including the assignment of identifiers to other "delegated identifier authority". A "policy" is a set of rules or conditions used by an authority to control interactions with a resource. "XDI account" represents a data authority hosted by an "XDI service provider" (XSP) acting as a delegated authority. "XDI link" is a data sharing relationship between two XDI resources. In addition, "XDI link contract" is XDI resource that controls the sharing of data across an XDI link. "XRI" is an extensible resource identifier, an URI-compatible abstract identifier. "XRI synonym" is any two or more XRIs that are asserted in an XDI document to be identifiers for the same XDI resource. Moreover, "root delegated data authority" (Root DDA) is the starting data authority in the delegation path established by an individual or organization that does not serve as its own XSP, but chooses to have an XDI account with another XSP. "XDI community" is considered a set of authorities that share a common XDI link contract governing their XDI data sharing relationships. XDI Document is an XML document conforming to the XDI meta-schema. Note that due to the proposed architecture of this meta-schema, XDI documents may be recursive to any depth. Finally an entity can be an individual, organization, or object that could represent a Web resource, a planet, etc. OASIS [50] defines eXtensible Resource Identifier (abbreviated XRI) is a scheme and resolution protocol for abstract identifiers that aim to provide a universal format for abstract, structured and platform-independent identifiers, so they can be shared across any number of domains, directories, and interaction protocols. The XRI specifications sit on top of the foundation provided by the Uniform Resource Identifier (URI) and Internationalized Resource Identifier (IRI) specifications published by IETF and W3C. XRI offers a lightweight resolution scheme using HTTP and simple XML documents. An XRI can contain another XRI to any level of nesting for cross-referencing purpose. This would enable defining structured identifiers that enable identifier sharing across domains in the same way does XML to enable data sharing across domains. In addition, XRI syntax supports peer-to-peer addressing that allows any two network nodes to assign to each other XRIs and perform cross-resolution. This helps in federating namespaces between organizations. The XRI resolution protocol includes a trusted version that uses SAML assertions.

Just as a URL is an address for a website, an I-name is an Internet address for the user. It is used to authenticate and share personal data. I-name technology is promising privacy; therefore, the user's identity won't be appearing to spammers and/or marketers without having the user to express its permission. When registering at the i-broker, the user gets back few services such as authentication service using offered through SAML-based i-Single Sign-On (i-SSO), contact page, forwarding service, i-link, i-mail, etc. I-names are human-friendly XRIs intended to be as easy as possible for people to remember and use. They are composed entirely of re-assignable segments as follows:

```
=firstname.lastname (individuals)
@company.and.Brothers (organizations)
+phone.number
=firstname.lastname / (+phone.number)
```

I-numbers are machine-friendly XRI that are registered to resources and never re-assigned. They are composed of entirely of persistent segments as follows:

```
!!1002!A8C9/!D50F.55
```

XRI resolution protocol can be used to resolve either i-name or i-number (or any combination of the two) [51]. An i-name is not "spamable" because it is not an email address (or a phone number, or a fax number, or any other form of direct communications channel.) Instead the owner of i-name controls how it

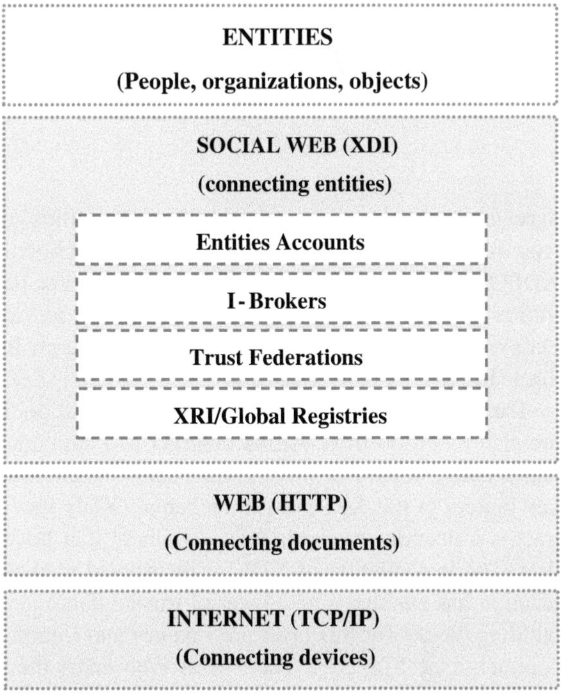

Fig. 3.5 The new layer of social Web

is resolved, and what privacy rules must be observed before any contact can be made or data accessed. This enables gateways that can automatically filter contact requests [52].

The DataWeb is an open source project that provides access services to statistical data like the current Web provides access services to documents. It defines a globally-distributed data sharing, which is based, as in the real-world, on social and legal contracts mechanisms that bind entities. Moreover, the DataWeb is based on the architectural style REST, which defines practices of Web services creation, in order to share and link digital data across domains and applications. Currently DataWeb is enabling social Web and services based on them are under development by XDI.ORG, an international non-profit organization. The term social Web has been first mentioned by Hoschka [53], Krey [54] and then, the members of the OASIS XDI Technical Committee have introduced it with XDI/XRI specifications in [52]. Based on the DataWeb definition, the social Web refers to an open global distributed data sharing network, which is considered as a part of Web 2.0. Instead of linking documents, the social Web links entities. The authors [55] make the analogy of social Web with the worldwide banking and credit card system since both of the systems are managing private and sensitive data: money and personal data. Like banks maintain accounts, i-brokers holds entities XRI accounts, Fig. 3.5. However, trust federation is a business alliance of i-brokers, who agree to abide by a common set of

Fig. 3.6 XDI and trusted data interchange

agreements in the care and handling of entities' data. A concrete example of trust federation is Identity Commons.[1] XRI/Global registries are monitored by XDI.ORG. Its mission is to provide community-based governance for the XRI global context registry and XDI data sharing services. The data browser for the DataWeb is Data Federated Electronic Research Review Extraction Tabulation Tool (FERRETT).

Trust is the foundation of any identity federation therefore; social Web and XDI are also based on trust. A link contract is a data control approach in a distributed data sharing network. Link contracts are fundamental to form the DataWeb and a key feature of the XRI Data Interchange (XDI) specifications. In XDI, a link contract is a machine-readable XDI document that governs the sharing of other XDI data. The introduction of XDI for distributed mediated data sharing and synchronization has enabled a new layer of trusted data interchange applications. The key building blocks for this layer are i-names and i-numbers, DataWeb pages, and link contracts (see Fig. 3.6). The Social Web takes the same approach for exchange of private, sensitive information by establishing a common means of exchange among trusted i-brokers [52, 56].

XDI links may exist between XDI resources under the control of a single Data Authority, or between different Data Authorities. XDI links within a single Data Authority may not require an XDI link contract, but XDI links between different data authorities will generally require an XDI link contract.

3.5 User-Centricity DigIdM Technical Models

Art Gilliland from Symantec points that user-centricity become important factor by questioning how do we make technology practical so that users can actually address their own privacy issues, their own auditing processes, and manage the protection of their data for themselves? [57] He adds that "if you look at the research that we've been doing, around 98 % of the data loss is through mistakes of human error and process breakdown. Being in the security industry, we're always going to be fighting the bad guys. But the bad guys are less of the problem around data loss. Being able to steal information is always going to be a business for somebody, and you can't ever fight all of them 100 %. But we can stop the large percentage that is human and process error" [22]. Patrick Heim,

[1] http://www.idcommons.net

the chief information security officer at Kaiser Permanente says: "We should not underestimate the human element. I liken it to driving. The reason we have controls in place such as driver's licenses is so that people at least have a basic understanding of the rules of the road and how to operate a vehicle safely, so that we can minimize those risks. I don't think there's been enough educational outreach to end users on how to use their systems safely. I'm not necessarily proposing there needs to be a "cyber driver's license," but you know, that probably wouldn't be a bad idea because we see that many, many of the observed problems are behavioral in nature" [57]. In the 54 page report [57] written by Sir James Crosby discusses how the UK can maximize the economic and social advantages of identity systems. The key element in common between the public and private sectors in managing identity is the consumer. The author stresses that these bodies should be moving to identity assurance focus rather than identity management one. The same author explains: "it is identity assurance that is best placed to meet a consumer's needs and to deliver mutual benefit to public and private sectors as well as to citizens" [57]. He adds: "the expression 'identity management' suggests data sharing and database consolidation, concepts which principally serve the interests of the owner of the database, for example the Government or the banks. Whereas we think of 'identity assurance' as a consumer-led concept, a process that meets an important consumer need without necessarily provides any spin-off benefits to the owner of any database. This distinction is fundamental. An identity system built primarily to deliver high levels of assurance for consumers and to command their trust has little in common with one inspired mainly by the ambitions of its owner" [57]. It is argued that it is user's identity, so he should be in the center of the process. He supports his argument about the importance of user-led identity management by pointing that the importance of identity systems goes beyond commercial transactions. He demonstrates that identity system will only help fulfill national security goals if it achieves mass usage. Thus, security objectives achievement lays on users' active participation [57]. He adds that user-center identity management and identity assurance are synonyms user-led identity management. Every aspect of an identity system should be designed from the consumer's perspective to realize the greatest economic and social benefits. The author suggests several principles on which should be laid any identity assurance system. Few of them include: (1) the purpose of any identity assurance system should be restricted to that of enabling users to assert their identity with ease and confidence; (2) the system governance should inspire the highest level of trust among citizens; (3) the amount of data should be minimized; (4) users should own their entry on any register; (5) enrollment process should be different for individuals with different circumstances, and change over time; (6) the system should be capable of being rolled out at pace to respond instantly to users' demands; (7) users, whose identity is compromised should be able to rely on their identity being repaired quickly and efficiently; and (8) enrollment and any token should be provided free of charge [57]. The authors say that in order to protect against more numerous and sophisticated attacks, security experts call for upgraded technology along with more attention

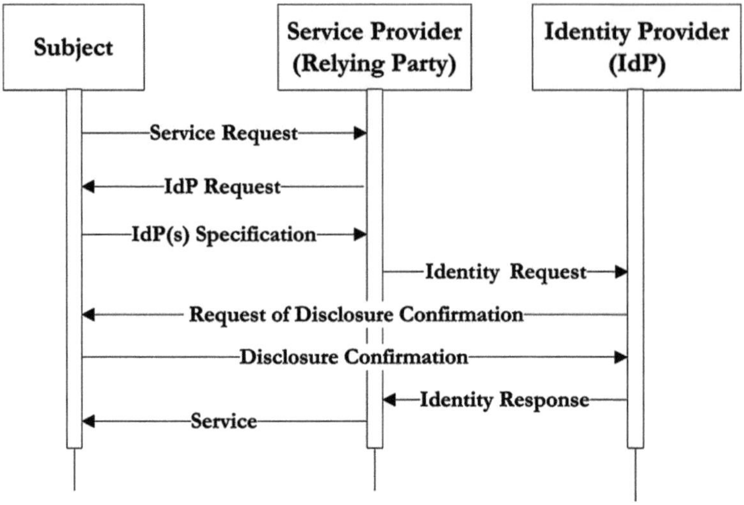

Fig. 3.7 User-centric identity federation

to human and legal factors [57]. In addition, Abhyankar, the senior director of product management at McAfee Avert Labs says: "The human element is something that we can't ignore" [57]. How users perceive the value of the service? Is it easy to use the service? Does the user feel that he has full control of his identity data? are critical questions that DigIdM systems designers should take into consideration when building such systems. We believe that the complexity of identity management comes from multiple reasons such as the nature of identity that has multiple facets such as technological, social, legal, and cultural; and the immaturity of digital identity and its related concepts in the digital life such as digital privacy. The system will likely be successful only if it balances added convenience with trust in the system [44]. User control over his attributes helps to build trust in identity systems.

Figure 3.7 illustrates the need of federated digital identity systems and within such system the subject still has the control over his attributes since he, and the only one, who could give to SP which IdP to contact. A typical identity federation conversation takes the form of two-way conversation between the subject, SP and IdP. Here is the conversation: (1) Subject → SP: Hello, I'd like Service A; (2) SP → IdP: I require attributes X, Y, and Z; (3) IdP → SP: here are the attributes X, Y, and Z; (4) SP → Subject: here is the Service A. A user-centric identity federation conversation, Fig. 3.7, is as follows: (1) Subject → SP: Hello, I'd like Service A; (2) SP → Subject: I require attributes X, Y, and Z, which IdP or IdPs should I ask? (3) Subject → SP: here is (are) IdP(s) to ask; (4) SP → IdP(s): I require attributes X, Y, and Z; (5) IdP(s) → Subject: do you confirm digital identity disclosure? (6) Subject → IdP(s): Yes, I confirm; (7) IdP(s) → SP: here are attributes X, Y, and Z; (8) SP → Subject: here is the Service A.

3.6 Making Less Visible Persistent Digital Identity

"Imagine being able to remember every fight you ever had with a friend, every time someone let you down, all the stupid mistakes you've ever made (…) I never forget anything, good or bad, so it is hard to move on", says Jill Price, a Californian woman who can't forget and has a brain power and a flawless memory to perfectly recall every trivial detail of every day's life without using mnemonic tricks. She provided exact dates of Elvis Presley death, plane crash in Chicago, Easters from 1980 to 2003, and even broadcast's date of Dallas TV series episode that revealed who shot J.R. Neurobiologists at the University of California-Irvine have coined a new name for her condition, calling it "hyperthymestic syndrome". The term has Greek roots, "thymesis" for remembering, and "hyper", meaning more than normal [58, 59].

3.6.1 Un-forgotten Digital Identity and Un-forgiven Digital Society

In his book [60], the author explores remembering and forgetting over human history and into the digital age. "Since the beginning of time, for us human, forgetting has been the norm and remembering the exception". But this balance has shifted, because of widespread of digital technology and global networks, forgetting has become the exception and remembering the default. The same author has also questioned on what are the potential consequences of this shift on both individual level and society level? What are the roles of forgetting and remembering in our society, and how these roles are changing? Is everyone who self-discloses information loses control over that information forever, and have no say about whether and when the Internet forgets this information? Do we want a future that is forever unforgiving because it is un-forgetting? He adds that the chilling effect of perfect memory alters our behavior [60]. The digital age is promoting the spread of hyperthymestic syndrome, which may yield to "un-forgiving" syndrome. The digital footprints that we leave on the Internet cannot be erased and fuel the "un-forgetting" memories. At anytime, footprints that are perfectly remembered by the Internet could be easily recalled and used against us. The author of the book [60] notes that perfect remembering make us un-forgiving to both ourselves and others, thus, he warns societies about such syndrome from which its consequences would be seen in the near future.

3.6.2 Digital Identity Persistence and Loss of Control

A 25-year-old single mother had completed her coursework and was looking forward to her future career in teaching. But she was denied her certificate, she was told because her behavior was unbecoming of a teacher. She had put a photo on MySpace showing her in costume wearing a pirate's hat and drinking from a plastic cup and captioned it "drunken pirate" for her friends to see. The university administration has argued that the photo was unprofessional. She considered taking the photo offline but the damage was done. Her web page had been catalogued by search engines and her photo archived by web crawlers. The Internet un-forgivingly remembered what Stacy wanted to have forgotten [60]. Therefore, Stacy has lost control her identity information, which implies irreversible and undesirable consequences: invasion of personal space, comfort and privacy, reputation harm, and openness to power abuses. This is a case to illustrate the un-forgiving consequence of un-forgetting memories.

Internet users are increasingly losing control over digital identity. They are leaving online trails when browsing the Web and disclosing more personal information, on which many service providers depend. Digital identities are considered as a row material for social-networking sites. Spock.com is offering people search engine services that would help to find people on the web and more specifically people who have profiles on social networks Live Spaces, Friendster, Hi5, MySpace, and Wikipedia. Spock's mission is to aggregate the world's people information and make it searchable. It is devoted to finding, indexing and profiling people on the Internet. Moreover, Spock provides to people tagging capabilities that could compromising reputations on the internet. Digital identities and user profiles allow to individuals accessing online services and for this reason they become valuable assets. Personal information can be found on websites and in publicly accessible databases. There is more than enough information for an unscrupulous criminal to take over people identity. Companies are using systems that analyze public records such as city's registry, credit files and the register of births, deaths and marriages to build a complete picture of a user online digital footprint. The systems can also analyze the content of social networks to build up a picture of the user relationship to other people. Companies are using applications of semantic tools, designed to bring meaning to large amounts of data [61].

3.6.3 Digital Identity Hiding and User Control

Personal data and security of identity information can be achieved by concept of identity hiding. Many tools, such as search engines, have been created to turn the Web into more visible and accessible platform but today users are requesting tools and features to have control over identity and particularly be less invisible. Web users are increasingly leaving trails on the net and most online service providers memorize, access and exploit 'Web of trails' for their own commercial benefits. As far service providers are processing identity information, as far users are losing control over their personal information that could compromise online security, privacy and trust [1, 9, 62, 63]. One hundred million worldwide Facebook users are threatened by identity theft, cyber-stalking and cyber-bullying, and digital espionage as a repercussion of Facebook hack case [64], in which personal details have been collated from public Facebook people directory and published through file-sharing service. Not only people are posting personal facts on the Web but government agencies are steadily making databases available online. The databases may include birth, marriage and death certificates, credit histories, voter registrations and property deeds [65]. Stacy's profile in social network made her identity more visible and lost ownership and control over it since she couldn't make her photo invisible or delete it. In addition, digitizing dossiers promoted identity loss of control and easy accessing them, which in turn encouraged identity theft and fraud. Currently, users feel concerned and worried about security, but providing control over identity would inevitably establish a community of trust and foster collaboration between business parties. "This tension between individuals' interest in protecting their privacy and companies' interest in exploiting personal information could be resolved by giving people more control. They could be given the right to see and correct the information about them that an organization holds, and to be told how it was used and with whom it was shared" [66]. In his book [60], the author argues that making identity information less visible, or giving "the right to be let alone" [67], is an efficient way to provide user's control and revive forgetting in un-forgetting digital identity. However, the author [68] argues that the word "trash" implies the remnants of something used but later discarded. It always contains traces and signatures of use such as monthly bills, receipts, personal papers, cellophane wrapping, price tags, and spoiled food. He stresses that future avant-garde practices will be those of trash and nonexistence, which is how does one develop techniques and technologies to make somebody unaccounted for? He illustrates with the example of laser pointer that can blind a surveillance camera when the beam is directed at the lens and as a consequence, the individual is not hiding but simply nonexistent to that node. We present an approach based on the use of metadata to make digital identity less visible and therefore gives the subject more control over it.

3.6.4 Digital Renaissance of Metadata

Metadata, information about information, called also "hidden data" [69] are being democratized and used for various purposes. From antiquity metadata have been created to codify knowledge and classify library materials in the goal to be more accessible. Assyrians attached small labels to clay tablets; Hittites and Ptolemies maintained catalogs of bibliographical entries and shelving information; and the library classification system in Chinese imperial library, Arabs 'halls of science', and renaissance's public libraries was based on bookmarking catalogues. As information become more abundant, the main problem is no longer finding it but accessing it easily and quickly. Today, by aiming to organize the world's information, Google is adding metadata e.g. indexes and PageRank scores when crawling and indexing Web pages. With the advent of Web 2.0, Web users tag web sites, documents, photos and videos helping to label unstructured information so it can be easily found through folkminds such as Delicious [70], Diigo [71], and [72]. Metadata is becoming a lucrative business opportunity since many companies and consumers are taking advantage of Amazon's popularity stars, bar codes and RFID labels. Photos uploaded to the website Flickr contain metadata such as when and often where they were taken, as well as the camera model, which could be useful for future buyers. As another example of metadata usage, MS Word document properties provided clues to police in order to resolve BTK killer case [63]. However, for any reason, such as for having fun or creating a buzz on the net, Web users could also instead labeling a photograph of a famous president as "president", they might bookmark it "terrorist" or "hacker". Thus, this phenomenon would have a negative side affecting people's digital identities and reputations [63, 69].

3.6.5 Metadata and Digital Identity Expiration Dates

We assume that digital identity is represented by either a single document or a set of documents (DigIdDoc) that comprises subject's attributes. Each digital identity document is linked with another document that comprises a set of metadata (DigIdMeta). XML-based DigIdMeta scheme comprises a set of beginning and ending tags classified into two sub-sets or document sections as shown in the following codes. The <Header> part, as shown in the following XML code, comprises all the tags that are related to DigIdDoc such as document identifier, name, disk location where it is saved, dates of creation, update, and disclosing. Other metadata related to DigIdDoc could be added such as the names of the person or machine that created, updated, and deleted the digital identity document. These metadata are useful for users to have more ownership or details about the owner, or owners, and dates of disclosure of his digital identity document [73].

```
<Header>
      <DocID>        </DocID>
      <DocName>      </DocName>
      <DocLocation> </DocLocation>
      <CreationDate> </CreationDate>
      <UpdateDate> </UpdateDate>
      <DisclosingDate> </DisclosingDate>
      <Names> </Names>
</Header>
```

The second part of the DigIdMeta code, as shown below, <PartyAgrt> comprises parties' agreements information as shown in the following code. It deals with restrictions, policies, rules and further legal requirements. Information about the discloser could be the expiration date of digital identity, which would reduce persistence and increase forgiving in societies. The discloser could be the person or a delegated and trusted party [73]. Author [60] explains et discusses the benefits and drawbacks of temporal dimension of information. DigIdMeta expiration date could revive forgetting by reducing digital identity age, thus reducing recall capabilities, in order to empower user control like disabling RFID chip. The recipient and permissible expiration date tags are represent the contract between disclosure and recipient or recipients. A negotiation process could be established in order to reach enough level of agreement upon min and max duration of expiration date with full alignment with and in accordance of permissible expiration date legal, policies, or rules requirements. And whether is it fixed or variable. This section could reduce "power issue" [60] and gives the user's more control over his digital identity.

```
<PartyAgrt>
      <Discloser>
            <ExpirationDate>  </ExpirationDate>
            <Visibility>      </Visibility>
      </Discloser>
      <Recipient>
            <ExpirationDateMin>     <ExpirationDateMin>
            <ExpirationDateMax>     <ExpirationDateMax>
            <Visibility>            </Visibility>
      </Recipient>
      <PermissibleExpDate>
            <Fixed>           </Fixed>
            <Min>             </Min>
            <Max>             </Max>
      </PermissibleExpDate>
      <Published>             </Published>
      ...
</PartyAgrt>
```

We can also add other information in this section such as whether this digital identity document is subject to aggregation, collection, and fusion or only limited to a specific purpose; purpose of retention in and delete from digital memories; digital identity accuracy rating that could be added only by the user or trusted party, a mean to challenge the quality of attributes' values. Such community powered tools are very popular to safe surfing the Web, e.g. recently Google has introduced WOT tool to rate Web sites. The owner of DigIdDoc could rate liability, credibility, and confidentiality of his document [73].

3.6.6 DigIdMeta and MetaEngine Tool

A MetaEngine will manage metadata and would help subjects to maintain a less visible digital identity. We present in Fig. 3.8, four environments where each is limited with an eclipse and behind each eclipse a subject. Links between eclipses represent an active and constant need of collaboration across different computing environments, such as operations of digital identity aggregation and profiling

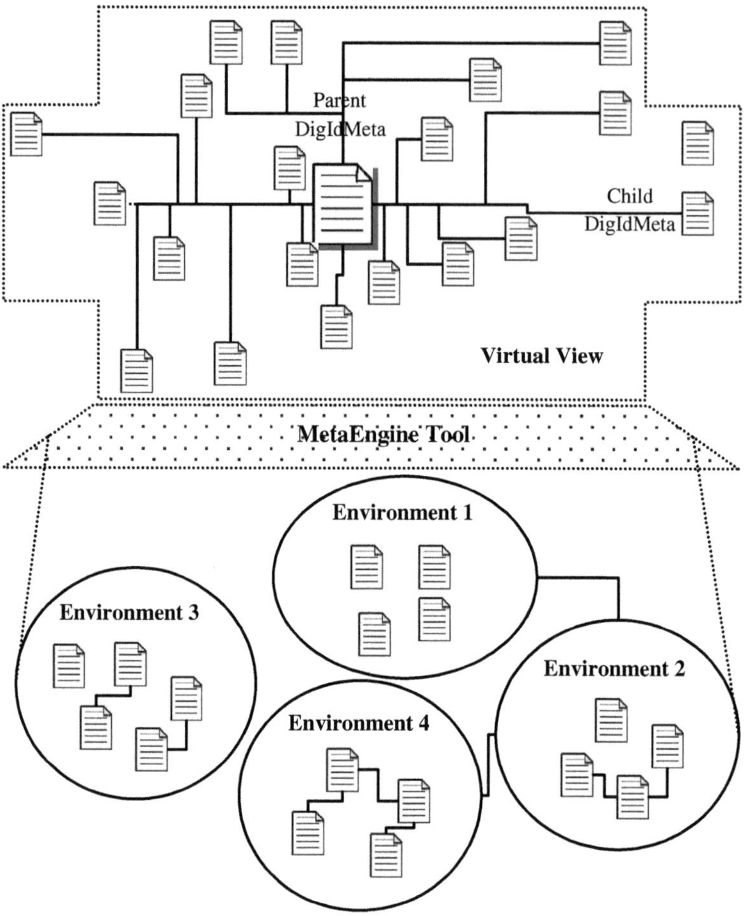

Fig. 3.8 DigIdMeta and MetaEngine tool

or a persistent link between two DigIdDocs residing in different ecosystems. Ecosystem 3 is isolated and is not liked to reflect a reality of a person who has a limited set of DigIdDocs. Such as a person who has limited activities using digital devices or striving to conduct anonymous activities. Documents residing inside the eclipses represent DigIdMeta documents attached to DigIdDocs, which are not represented in Fig. 3.8. At the intra-environment level, the composite DigIdMeta could be linked to each others; a subset of them is linked; or not linked. The link between DigIdMeta represents the link between digital identity documents that the subject has established. The link between two or more digital identity documents could represent: (a) the use of the same subject's account to access two or more services such as Google mail and YouTube. In this case, two distinct digital identity documents are created and generated comprising the same attributes but with different identifiers; (b) attributes' values change and evolution over time such as subject's interests, home address, and employer. In this case, a new digital identity document is created comprising attributes' new values and linked to the former document; and (c) a replication of static attribute in other digital identity documents, such as replication of ID number in driver's license document [74, 75].

Brain's forgetting mechanism is inspiring research on making digital identity less visible. Researchers are closely studying how the brain forgets information that is stored in long-term memory. Some think that when we forget means that we have lost the link to that information like Web pages URLs. Others reckon and suggest that our brain constantly reconfigures our memory and they say that what we remember is based, at least in part, on our present preferences and needs. Empirical research seems to support the second ideas [60]. Both ideas inspired us to consider adopting an engine that will provide DigIdDoc search, synchronizing, and refresh capabilities. The engine functions could remind a rubber bulb of blood pressure sphygmomanometer. Instead of pushing/pulling air, it will pull DigIdMeta documents from multiple data sources and push them to computing ecosystem's requester. As a result, the latter would receive a specific number of DigIdDocs ordered on a priority basis like any keyword search engine result. The DigIdDoc priority order is calculated based on the weight_score, which is an output of the function, that combines two other scores: grain_score and distance_score [74], as follows.

```
Function WeightScore (input grain_score,
distance_score): output weight_score.
```

Whenever a computing ecosystem requests a subject's digital identity, MetaEngine will collect all DigIdMeta associated with subject's DigIdDocs and push them into a virtual view. This is similar to data aggregation conducted via virtual directory in which collected data are maintained within non physical settings and the virtual view disappears whenever the operation is no longer needed. The collected DigIdMeta are shown inside the discontinuing-line shape. Besides, MetaEngine tool will calculate the grain score for each DigIdDoc, write it in its DigIdMeta and elect the one that has the highest score to be the parent, or top-level, document, a shadowed one in Fig. 3.8. The parent DigIdMeta will be located in the center and surrounded by other child DigIdMeta. This is like a fact table in a data warehouse's star data schema, which is surrounded by dimension tables. Moreover, the MetaEngine tool will include all the links to the surrounded children in the parent's DigIdMeta and the distance score of each link in the child's DigIdMeta. MetaEngine invokes the function WeightScore to calculate the weight_scores and writes each weight_score in its associated child DigIdMeta. The parent DigIdMeta has neither a distance score nor a weight. It has the highest grain_score and the associated DigIdDoc will appear in the top of the search ordered list like a search engine result. Each of the following

DigIdDocs on the list will be ordered on the basis of how high the weight_score. The distance_score would empower the "forgetting" capabilities. MetaEngine tool would make a specific number of DigIdDocs, which have higher distance_scores, easy to access comparing to the ones that have a lower distance_scores. The latter should be hard to retrieve and to be accessed. For instance, low distance_score will be on the bottom of search result list, the disclosing decision is followed by the subject's communication of his consent, or the ecosystem should request many times in order to access distant DigIdDocs. MetaEngine tool conducts the refresh operation on on-demand basis, whenever the requester asks for DigIdDocs. It aggregates DigIdMeta, synchronizes the duplicates, recalculates the scores, and reorganizes the links. In the following subsections, we present few parameters that could be used to calculate GrainScore and DistanceScore. We do not intend to provide functions' parameters but we present few clouts that could have a direct or indirect impact on the scores [74]. Work in this area is still in progress.

The central DigIdDoc is the document that has the highest relevance score. The grain_score is to be calculated on the basis of a set of parameters such as activity and popularity rates. Activity rate represent how actively the subject is using the digital identity document. For instance, the subject could be using frequently the Gmail profile/account more than the Yahoo one, thus the activity rate of the latter is lower than Gmail profile. Popularity rate represents how others perceive subject's identity such as a number of user's tags, a number of users' generated bookmarks on a subject's web page, a number of comments in personal blogs, and a number of blogroll links that point the subject's blog [74]. Distance_score is calculated based on multiple criteria. For instance, DigIdDoc expiration date could be set by the subject, by computing ecosystem's service provider, or dictated by law. In addition, we can consider forgetting probability and elapsed time from DigIdDoc creation date. As much the distance_score is higher as far is the child DigIdMeta from the parent one [74]. Cooling functions (i.e. forgetting functions) model the apparent loss of information memorized by a human brain for machine computing. It is important to notice that the human brain tends to forget not because it has a limited capacity memory but rather information units tends to interfere with each other and be aggregated in a way that older information units become more and more inaccessible. We identify two parameters of the distance function: elapsed time t and a random vector, which is defined by joint probability density for forgetting between two documents [75]. We present, below, an overview of the XRD document structure and an implementation of DigIdMeta document. Recently published as an OASIS standard, XRD is a simple generic format for describing resources. XRD documents provide machine-readable information about resources for the purpose of promoting interoperability, which is an important need for collaboration across systems. The following XML schema fragment defines the XML namespaces, location of the normative XML Schema file for an XRD document and other header information for the XRD schema [76].

```
<schema targetNamespace="http://docs.oasis-
open.org/ns/xri/xrd-1.0"
  xmlns="http://www.w3.org/2001/XMLSchema"
  xmlns:xrd="http://docs.oasis-open.org/ns/xri/xrd-1.0"
  xmlns:ds="http://www.w3.org/2000/09/xmldsig#"
  elementFormDefault="unqualified"
  attributeFormDefault="unqualified"
  blockDefault="substitution"
  version="1.0">
<import namespace="http://www.w3.org/2000/09/xmldsig#"
schemaLocation="http://www.w3.org/TR/2002/REC-xmldsig-
core-20020212/xmldsig-core-schema.xsd"/>
<import namespace="http://www.w3.org/XML/1998/namespace"
        schemaLocation="http://www.w3.org/2001/xml.xsd"/>
<annotation>
  <documentation>
     Document identifier: xrd-schema-1.0
     Location: http://docs.oasis-open.org/xri/xrd/v1.0/
  </documentation>
</annotation>
...
</schema>
```

XRD provides XML format for describing meta-documents. XRD DigIdMeta document, Fig. 3.9, describes properties of the document itself, as well as the relationships with other DigIdMeta documents.

XRD DigIdMeta document can be divided into two main sections, Fig. 3.9: (1) document header section that includes a description of the XRD DigIdMeta document itself, such as document's expiration date [60], and XML namespaces; and (2) resource information section, which is divided into two subsections: resource's description and resource's associated links. The document's description subsection includes properties and aliases of the DigIdDoc, and the next subsection lists links to other DigIdDocs. If a requester's ecosystem wants to know and learn more about the DigIdDoc, identified by an URI, it retrieves its XRD DigIdMeta document. XRD DigIdMeta provides characteristics and attributes enclosed between <property> tags; and the relationships to other DigIdDocs and available associated services within <links> tags. XRD DigIdMeta document is bounded to DigIdDoc through either the unique identifier URI or an alias, which is an alternative and human-friendly URI. The <Expires> element defines XRD DigIdMeta document life duration, which could be set by the developer and/or HTTP protocol. The element <property> describes the digital identity document with URI-formatted strings. Finally, XRD DigIdMeta document encapsulates links to other DigIdDocs between <link> tags [76, 77].We present above the XRD implementation of resource information section of the DigIdMeta document. The value between <subject> tags is the unique identifier of the document.

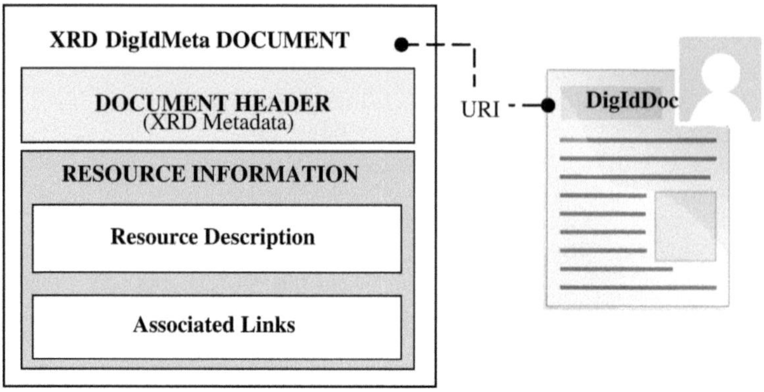

Fig. 3.9 DigIdDoc and XRD DigIdMeta

Multiple <aliases> could be included to have contextual identifiers and avoiding unique and universal identifier, which could harm privacy. Disk location is enclosed as a property to know the locations of DigIdDoc and its related DigIdMeta document. If the duplicate's value is set to "Yes" then links to duplicated DigIdDocs are to be added. Subject's DigIdDoc expiration date, recipient's minimum and maximum expiration dates, and/or legally permissible expiration dates could be either considered as properties in XRD DigIdMeta or as input variables in DistanceScore function. Multiple disclosing dates could be added into the DigIdMeta to ensure a tracking of a few least disclosures. Links to DigIdMeta children are configured by MetaEngine during pulling/pushing operations. DigIdMeta links could add consistency in DigIdDocs search operation and this could be a mean to overcome identity resolution issues associated with having many people with the same full-name [74].

```
<XRD>
<Subject>http://www.favorite-social.net/gba</Subject>
<Alias> http://www.favorite-socialnet.net/ghazi.benayed
</Alias>
<Alias> http://www.favorite-socialnet.net/ghazibenayed
</Alias>
<Expires>XRD_expiration_date_value</Expires>
<Property type='http://favorite-
social.net/gba/expDate'>DigIdDoc_expiration_date_value</Pr
operty>
<Property type='http://favorite-
social.net/gba/location'>DigIdDoc_location</Property>
<Property type='http://favorite-
social.net/gba/duplicate'>Y/N </Property>

// This section is bounded to child's document
<Property type='http://favorite-social.net/gba/gs'>
grain_score_value </Property>
<Property type='http://favorite-social.net/gba/ds'>
distance_score_value </Property>
<Property type='http://favorite-social.net/gba/ws'>
```

```
weight_score_value </Property>
<Property type='http://favorite-social.net/gba/cd'>
creation_date</Property>

<Property type='http://favorite-social.net/gba/dd'>
last_disclosing_date </Property>
<Property type='http://favorite-social.net/gba/dexpd'>
discloser_expiration_date </Property>
<Property type='http://favorite-social.net/gba/minrexpd'>
min_discloser_expiration_date </Property>
<Property type='http://favorite-social.net/gba/maxrexpd'>
max_discloser_expiration_date </Property>

// The Link section is bounded to parent's document
<Link rel='update' type='text/html'
         href='http://favorite-social.net/gba/update'>
<Title xml:lang='en-us'>Link to Updated DigIdDoc </Title>
</Link>
<Link rel='duplicate' type='text/html'
         href='http://favorite-social.net/gba/duplicate'>
<Title xml:lang='en-us'>Link to Duplicated DigIdDoc
</Title> </Link>
<Link rel='child1' type='text/html'
         href='http://favorite-social.net/gba/child1'>
<Title xml:lang='en-us'>Link to Child1 DigIdDoc </Title>
</Link>
...
<Link rel='childn' type='text/html'
         href='http://favorite-social.net/gba/childn'>
<Title xml:lang='en-us'>Link to Childn DigIdDoc </Title>
</Link>
</XRD>
```

3.6.7 Expiration Date Within Content-Centric Network

Evolving from a document-centered into a service and data-centered World Wide Web, Web of data, requires a better user's digital identity protection and management. The permanence nature of digital identity entails loss of user's control over distributed identity attributes and privacy breaches. We propose an innovative Stop-Dissemination mechanism that is built on the basis of data expiration date techniques coupled within the promising Content Centric Network capabilities. Two use cases are detailed to explain the mechanism in order to have low permanence of federated digital identity documents [78].

Currently, the use of the internet has changed from machine interconnection to data and service oriented communication. As a consequence, data centric infrastructures and architectures are proposed to spin off Internet from simple host to host communication model into data delivery and manipulation. IP address is no longer a key identifier; however, every piece of data is identified by a unique key, called a content name. New data delivery mechanism is based on two elements: (1) data naming is the content name attribution process; and (2) name resolution is a locating process to find the appropriate host that holds a valid copy of the requested data [79–82]. CCN is one of the recent projects in the data centric internetworking field. It offers new naming and resolution mechanisms. CCN names are built hierarchically from specified components. The name is composed from at least: a globally routable name and organizational name. CCN relies on two packets to perform name resolution and data delivery: (1) interest packet is broadcasted by a consumer over all the possible and available connectivity to express his interest in a specific content; and (2) data packet responds to requests [83]. CCN's content refers to data and the equivalent of Internet IP router is the forwarding engine. A consumer asks for a specific content by issuing an interest packet that encloses a content name and extra options such as data filter and order preference. The consumer sends the interest packet to the nearest CCN forwarding engine, which is in-charge of the name resolution and data delivery. The forwarding engine has basically three tables: (1) the Forwarding Information Base (FIB) is employed to forward interest packet to eventual sources; (2) Content Store is a buffer memory that stores data packets, which have pre-established replacement policies. For instance, an administrator can choose short packet life-time to quickly recycle the buffer or a long life-time to serve more consumers; and (3) Pending Interest Table (PIT) keeps track of forwarded interest packet to be able to send returned data to its requestors. After the reception of the interest packet, the forwarding engine performs a lookup in the following order: It searches in the content store, then in the PIT and the FIB. If there's a data packet in the content store that matches the interest, it will be sent immediately to the requestor. But, if there's no match in the content store, the engine will search in the PIT that stores on-going requests. If the match is conducted, the engine will not forward the current interest packet. Simply, it adds the requestor in the "Requesting Faces List" and the packet is discarded. If content store and PIT don't satisfy the request, the engine looks in FIB. If the engine finds a matching source, it will forward the interest to that source and creates a new entry in PIT. Lastly, if within three attempts no matching solution is found, the engine discards the interest packet [83].

In opposition to other data-centric internetworking infrastructures such as DONA, PSIRP, NetInf, and DHT-Based Solutions, we choose a CCN infrastructure for several reasons: (1) CCN name resolution is the nearest approach to the current Internet infrastructure. Thus no big changes are required; (2) CCN infrastructure is very close to that of the Internet and it offers data recognition capabilities. CCN could provide a better way of DigIdDocs management, which may yield to a better digital identity protection level and control; (3) CCN offers flexibility in customizing the networking communication model; (4) CCN allows managing multiple data types, which respond to the need of ubiquity and digital life in which different types of identity documents (DigIdDocs) are to be created and shared; (5) Subjects are more and more delegating the task of digital identity management to application software, which should be well designed to ensure the protection of digital identity attributes. Security aspects should be taken into consideration from the outset design of such systems and therefore digital identity security and protection costs have to be supported by the subjects [78].

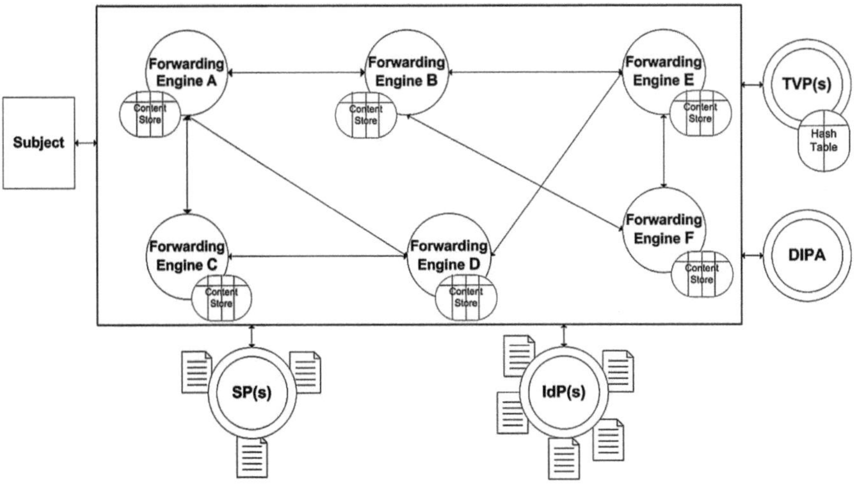

Fig. 3.10 CCN-based digital identity federation

However, CCN infrastructures could offer preconfigured mechanisms of DigIdDocs management that would necessary reduce DigIdDocs security and protection system engineering costs. DigIdDoc within CCN infrastructure is identified by a unique content name and it is considered as any content that could be exchanged between participants [78].

Federated identity systems are medium that allows collaboration between participants within a circle-of-trust. Basic federated architecture involves multiple participants: (1) Subject could represents an individual, a software component, or a computer; (2) Identity Provider (IdP) that can be a single or multiple providers [84]. Multiple documents surround the IdP to represent various subjects' DigIdDocs that IdP manages; (3) Service Providers (SP) which can be a single or plural providers of services such as ecommerce web site or email account. SP is surrounded by a limited number of documents to explain that such provider maintains only DigIdDocs of the subjects that have asked him for a service (see Fig. 3.10); and (4) other parties such as Trust Verification Provider (TVP) and Digital Identity Protection Authority (DIPA). TVP keeps a hash table. The rectangle drawn in Fig. 3.10 represents infrastructure's delimiter. It represents federation's circle of trust in which inside all interconnected forwarding engines and federation's participants are hooked up. Every CCN forwarding engine maintains a content store, which is a memory buffer that keeps multiple contents in form of data packets. The links between forwarding engines symbolize a two-way networking communication model [78].

The forwarding engine comprises a content store. The content store is composed of two columns: (1) CCN content names; and (2) data. Data in this context refers to a set of digital data packets. CCN provides a basic data packet that it's composed of header and data. Header encloses the content name and other

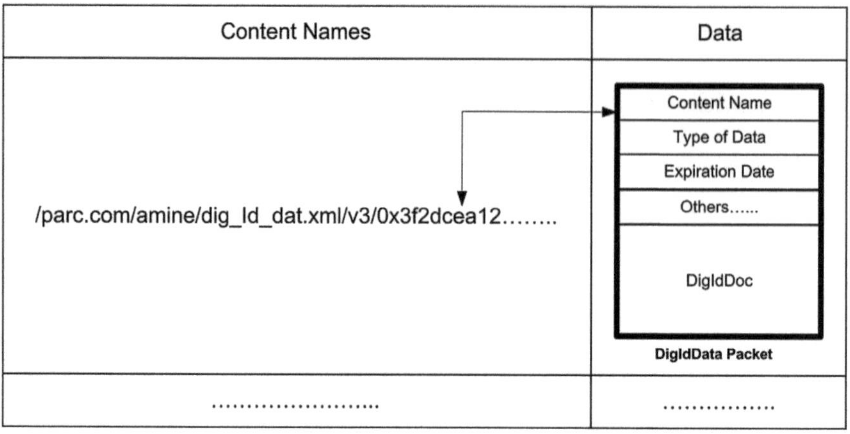

Fig. 3.11 CCN content store and DigIdData packet

name resolution information. We propose a new DigIdData packet (see Fig. 3.11). Beside the basic CCN data packet information, we propose to extend header section with two new fields: (1) content type refers to the multiple types of data that CCN infrastructure could support. "DigIdDocType" is the new value of the content type field that we propose when referring to DigIdDoc; and (2) expiration date or temporal dimension of identity attributes that could revive "forgetting" capabilities and reduce recall capabilities in the digital age. This is similar to disabling RFID chip [78].

Future avant-garde practices and techniques of Trash or nonexistence are encouraged to make oneself unaccounted for. A simple laser pointer can blind a surveillance camera when the beam is directed at the lens. In consequence, one is not hiding, simply nonexistent to that node [68]. Expiration date has been a medium to make digital identity less visible in various related work [60, 73] and an XRD implementation of the metadata containing expiration date is also suggested by Ben Ayed and Ghernaouti-Hélie [74]. Furthermore, data section encapsulates DigIdDoc. We describe two use cases in order to fully explain the mechanism that takes in place during the collaboration between participants over CCN core.

The use cases are: (1) service request use case (Fig. 3.12). A subject requests a service from a SP which in turn demands identity information from the subject. The subject sends IdP's information to the SP, which requests subject's identity information from the appropriate IdP. The latter invokes a Hash Calculation Method (HCM) that is based on a specific function requiring three input parameters to generate a unique hash key. The parameters are content name of DigIdData packet, DigIdDoc's expiration date, and SP CCN identifier. The IdP sends the hash key to TVP and the DigIdDoc to the SP. Finally, the SP sends access to the service to the subject. TVP is identified and added as a new party to the system in order to provide hash verification. The hash key will be used later to limit DigIdDoc dissemination by the SP; (2) DigIdDoc stop-dissemination use case (Fig. 3.13). An opportunist SP

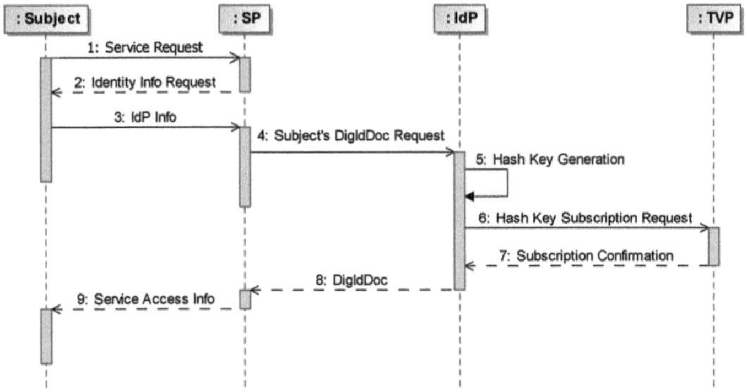

Fig. 3.12 Sequence diagram of service request use case

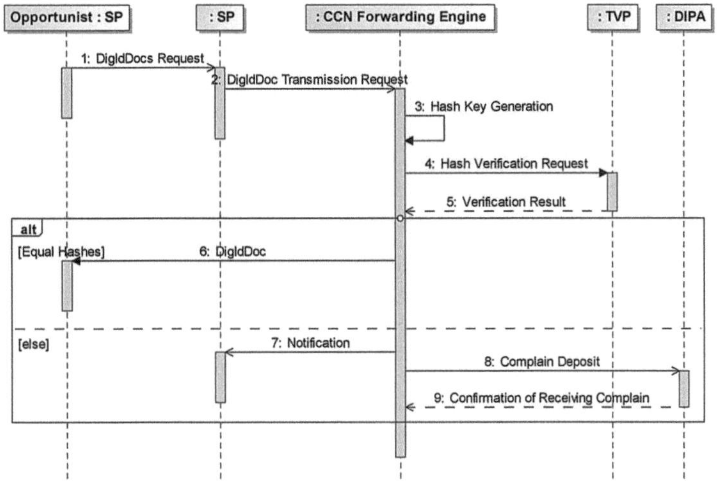

Fig. 3.13 Sequence diagram of DigIdDoc stop-dissemination use case

could send a DigIdDocs request to the SP trying to take advantage of the available digital identities by collecting, analyzing, processing DigIdDocs for commercial purposes and for other purposes rather than the original one. The SP asks a forwarding engine to transmit DigIdData packet, which encapsulates a DigIdDoc, to the opportunist SP. The forwarding engine checks the content type of the packet. When it is a DigIdData, the forwarding engine generates a hash key calling the same HCM. The Hash key is sent to the TVP, which checks whether it exists in the hash table. If it is, the forwarding engine elicits the transmission eligibility. If it is not, the forwarding engine deduces that the DigIdData packet is illegally in use and a fraud notification is sent right away to a certain policy/legal authority DIPA. The interface to the authority's system allows an easy complaint deposit [78].

References

1. P.J. Windley, Digital identity: Unmasking identity management architecture (IMA). O'Reilly Media (2005)
2. A survey. 01 Informatique magazine (2004)
3. Organizing Committee of Digital Identity & Privacy (Human Capital & Social Innovation Technology Summit), Call for contribution to managing digital identities for education, employment and business development (2007), Available: http://events.eife-l.org/HCSIT2007/overview/dip/dip2007. Accessed 11 May 2010
4. International Telecommunication Union—Joint Coordination Activity for Identity Management (JCA-IdM), Scope of identity management, Available: http://www.itu.int/en/ITU-T/jca/idm/Pages/default.aspx. Accessed 11 Feb 2010
5. S. Vanamali, Identity management framework. Inf. Syst. Control J. **4** (2004)
6. D.G.W. Birch, The identity vision, in *Digital Identity Management: Perspectives on the Technological, Business and Social Implications,* ed. by D.G.W. Birch (Gower Publishing Limited, 2007), pp. 3–8
7. M. Benantar, *Access Control Systems: Security, Identity Management and Trust Models* (Springer Science + Business Media, 2006)
8. G. Ben Ayed, Consolidating fragmented identity: Attributes aggregation to secure information systems. IADIS Int. J. Comput. Sci. Inf. Syst. **4**, 1–12 (2009)
9. International Telecommunication Union, Digital Life, ITU Internet Report (2006), Available: http://www.itu.int/osg/spu/publications/digitalife/docs/digital-life-web.pdf. Accessed 21 May 2010
10. W. Hommel, Using XACML for privacy control in SAML-based identity federations, in 9th IFIP TC-6 TC-11 Conference on Communications and Multimedia Security, Salzburg, Austria (2005), pp. 160–169
11. S. Slone, Identity management (2004), Available: http://www.opengroup.org/onlinepubs/7699959899/toc.pdf. Accessed 21 May 2010
12. A. Scorer, Identity directories and databases, in *Digital Identity Management: Perspectives on the Technological, Business and Social Implications,* ed. by D.G.W. Birch (Gower Publishing Limited, 2007), pp. 41–49
13. L. McRae, Identity management strategic vision (2005), Available: http://www.stanford.edu/dept/its/vision/identity-management.html. Accessed 4 May 2008
14. International Telecommunication Union—Focus Group on Identity Management (FG IdM), Report on identity management ecosystem and Lexion (2007)
15. E. Damiani et al., Managing multiple and dependable identities. IEEE Internet Comput. IEEE Comput. Soc. 29–37 (2003)
16. D.A. Buell, R. Sandhu, Identity management. IEEE Internet Comput. 26–28 (2003)
17. M. Small, Business and technical motivation for identity management. Inf Secure Tech R **9**, 6–21 (2004)
18. R.R. Panko, *Corporate Computer and Network Security*, 2nd edn. (Prentice Hall, New Jersey, 2009)
19. Organization for Economic Co-operation and Development (OECD), At crossroads: Personhood and digital identity in the information society (2008). The working paper series of the OECD directorate for science, technology and industry, Available: http://www.oecd.org/LongAbstract/0,3425,en_2649_34223_40204774_119684_1_1_1,00.html. Accessed 21 May 2010
20. H. Chivers, Personal Attributes and Privacy: How to ensure that private attributes management is not subverted by data mining (2004), Available: http://sec.cs.kent.ac.uk/cms2004/Program/CMS2004final/p1a2.pdf. Accessed 24 March 2011
21. M. Hansen, User-controlled identity management: The key to the future of privacy? Int. J. Intellect. Property Manag. **2**, 325–344 (2008)

22. J. Crosby, Challenges and opportunities in identity assurance (2008)
23. PriceWaterhouseCoopers Thought Leadership Institute, From the white board to the bottom line: The case for pursuing process maturity through business process management (2010), Available: http://download.pwc.com/ie/pubs/from_the_white_board_to_the_bottom_line.pdf. Accessed 18 March 2011
24. Sxip, Sxip, Available: http://www.sxip.org. Accessed 11 Feb 2010
25. LID, LID, Available: http://lid.netmesh.org/wiki/Main_Page. Accessed 11 Feb 2010
26. XDI/XRI, XDI/XRI, Available: http://www.xdi.org/. Accessed 11 Feb 2010
27. OpenID, OpenID, Available: http://openid.net/. Accessed 11 Feb 2010
28. Yadis, Yadis, Available: http://yadis.org/wiki/Main_Page. Accessed 11 Feb 2010
29. Microsoft, Windows CardSpace, Available: http://www.microsoft.com/windows/products/winfamily/cardspace/default.mspx. Accessed 11 Feb 2010
30. Telecommunication Standardization Sector (ITU-T), Available: http://www.itu.int/ITU-T/. Accessed 11 Feb 2010
31. The Internet Engineering Task Force (IETF), Available: http://www.ietf.org/. Accessed 11 Feb 2010
32. Kantara Initiative, Available: http://kantarainitiative.org/. Accessed 11 Feb 2010
33. Liberty Alliance Project, Available: http://www.projectliberty.org/. Accessed 11 Feb 2010
34. Shibboleth, Shibboleth Project, Available: http://shibboleth.internet2.edu/. Accessed 14 Feb 2010
35. Bandit Project, Available: http://www.bandit-project.org/. Accessed 8 May 2008
36. Higgins Community, Higgins Open Source Identity Framework, Available: http://eclipse.org/higgins/. Accessed 14 Feb 2010
37. PRIME Community, PRIME—Privacy and identity management for Europe document, Available: https://www.prime-project.eu/. Accessed 14 Feb 2010
38. International Telecommunication Union—Focus Group on Identity Management (FG IdM), Report on identity management use cases and gap analysis (2008)
39. R. Oppliger, Microsoft.NET passport and identity management. Inf. Secur. Tech. Rep. **9**, 26–34 (2004)
40. K.-K.R. Choo, Issue report on business adoption of microsoft passport. Inf. Manag. Comput. Secur. **14**, 218–234 (2006)
41. M. Gardiner, The business value of identity federation (2007), Available: http://whitepaper.techworld.com/authentication/4818/the-business-value-of-identity-federation. Accessed 21 May 2010
42. J. Hodges et al., Glossary for the OASIS security assertion markup language (SAML) V2.0, OASIS (2005)
43. J. Hodges, Liberty technical glossary, Liberty Alliance Project (2006)
44. Center for Democracy & Technology, Privacy Principles for Identity in the Digital Age [Draft for Comment—Version 1.4] (2007), Available: http://www.cdt.org/files/pdfs/20071201_IDPrivacyPrinciples.pdf. Accessed 28 May 2010
45. B. Pfitzmann, Federated identity-management protocols (Transcript of Discussion). LNCS J. (Springer, 2005), pp. 175–177
46. N. Klingenstein, Attribute aggregation and federated identity, in *Proceedings of the IEEE International Symposium on Applications and the Internet Workshops* (2007)
47. Q. Pham et al., Consistency of user attribute in federated systems. LNCS J. (Springer, 2007)
48. OASIS, XRI requirements and glossary (2003)
49. OASIS, Extensible resource identifier (XRI) resolution (2006)
50. OASIS, An introduction to XRIs (Working Draft) (2005)
51. XDI, XDI, Available: http://www.xdi.org/. Accessed 21st May 2010
52. D. Reed et al., The social web: Creating an open social network with XDI. PlanetWork J. (2004)
53. P. Hoschka, CSCW research at GMD-FIT: From basic groupware to the social Web. ACM SIGGROUP Bull. **19**, 5–9 (1998)
54. A.C. Krey, History and the social Web (University of Minnesota Press, 1995)

55. J. Caroll, J. Murphy, Who am I? I am Me! identity management in a networked world, in *Proceedings of the 4th International We-B Conference* (2003)
56. The Dataweb: An introduction to XDI, Available: http://www.oasis-open.org/committees/download.php/6434/wd-xdi-intro-white-paper-2004-04-12.pdf. Accessed 21 May 2010
57. R. Abhyankar et al., (2008) Improving online security. Sci. Am. Mag. 96–99
58. S. Kruglinski, The woman who never forgets: Does AJ have the world's best memory? DISCOVER Mag. (2006), Available: http://discovermagazine.com/2006/jun/j-woman-memory. Accessed 25 Nov 2010
59. J. Adler, Unable to forget: The remarkable story of a woman who remembers every day of her life. Newsweek Mag. (2008), Available: http://www.newsweek.com/2008/05/10/unable-to-forget.html. Accessed 9 Jan 2011
60. V. Mayer-Schönberger, Delete: The virtue of forgetting in the digital age (Princeton University Press, 2009)
61. J. Fildes, Taking control of your digital ID (2006). Available: http://news.bbc.co.uk/2/hi/technology/6102694.stm. Accessed 22 May 2010
62. G. Bell, J. Gemmel, A digital life. Sci. Am. Mag. 58–65 (2007)
63. K. Cukier, A special report on managing information. The Economist (23rd Feb–5th March 2010)
64. Facebook 'hack' releases 100 million user details onto filesharing sites. Inforsecurity USA (2010), Available: http://www.infosecurity-us.com/view/11343/facebook-hack-releases-100-million-user-details-onto-filesharing-sites/. Accessed 30 Nov 2010
65. M. Fischetti, Scoring your identity: New tactics root out the false use of personal data. Sci. Am. 27–28 (2007)
66. K. Cukier, New Rules for Big Data: Regulators are having to rethink their brief. The Economist (23rd Feb–5th March 2010), Available: http://www.economist.com/specialreports/displaystory.cfm?story_id=15557487. Accessed 13 May 2010
67. P. Brown, Privacy in an Age of Terabytes and Terror. Sci. Am. Mag. 46–47 (2008)
68. A.R. Galloway, E. Thacker, The exploit—A theory of networks (University of Minnesota Press, 2008)
69. S.L. Garfinkel, Information of the World, UNITE! Sci. Am. Mag. 82–87 (2008)
70. Delicious, Available: http://www.delicious.com
71. Diigo, Available: http://www.diigo.com
72. Technorati, Available: http://technorati.com
73. G. Ben Ayed, Digital identity metadata scheme: A technical approach to reduce digital identity risks, in International Workshop on Information Security and Risk Management of the 25th IEEE International Conference on Advanced Information Networking and Applications (AINA-2011), Biopolis, Singapore (2011)
74. G. Ben Ayed, S. Ghernaouti-Hélie, XRD digital identity metadata-based approach to foster collaborations across networked computing ecosystems, in The Third International Conference on Networked Digital Technologies (NDT 2011), Macau, China (2011)
75. G. Ben Ayed et al., Towards building weak links between persistent digital identity documents: MetaEngine and distance to make identity less visible, in International Conference on Digital Enterprise and Information Systems (DEIS2011), London, UK (2011)
76. OASIS eXtensible Resource Identifier (XRI) TC, Extensible Resource Descriptor (XRD) Version 1.0 (OASIS Standard) (2010), Available: http://docs.oasis-open.org/xri/xrd/v1.0/xrd-1.0.html. Accessed 8 Dec 2010
77. E. Hammer-Lahav, XRD document structure (2009), Available: http://hueniverse.com/2009/03/xrd-document-structure/. Accessed 9 Dec (2010)
78. A. Elabidi et al., Towards hiding federated digital identity: Stop-dissemination mechanism in content-centric networking, in The 4th International Conference on Security of Information and Networks (SIN 2011), Sydney, Australia (2011)
79. D. Meyer et al., Report from the IAB Workshop on Routing and Addressing (RFC 4984) (2007)

80. D. Clark et al., Addressing reality: An architectural response to real world demands on the evolving internet, in ACM SIGCOMM Conference—Workshop on future directions in network architecture (FDNA-03), Germany (2003)
81. M. Handley, A. Greenhalgh, Steps towards a Dos-Resistant internet architecture, in ACM SIGCOMM Conference—Workshop on future directions in network architecture (FDNA-03), USA (2004)
82. V. Jacobson, If a Clean Slate is the solution what was the problem, in Stanford Clean Slate Seminar (2006)
83. V. Jacobson et al., Networking named content, in The 5th International Conference on Emerging Networking Experiments and Technologies (ACM CoNEXT '09), pp. 1–12 (2009)
84. Organisation for Economic Co-operation and Development, The role of digital identity management in the internet economy: A primer for policy makers (2009), Available: http://www.oecd.org/dataoecd/55/48/43091476.pdf. Accessed 16 June 2010

Chapter 4
Privacy and Digital Identity

Civilization is the progress toward a society of privacy.
Ayn Rand (Writer and Novelist, 1905–1982)

Privacy is a human right and an important need for societies to progress. It is considered as a requirement for maintaining the human condition with dignity and respect [1]. Interpreted broadly, privacy is about the integrity of the individual [2, 3] and an integral part of the his dignity [4]. In reviewing the literature, we noticed that privacy is a complex and subjective concept that has with different meanings to different people when used in different contexts. Therefore, it is currently used to refer to some quite specific needs or expectations of today's society such as freedom from the attention of paparazzi and protections against digital camera voyeurism. One of the most common narrow usages of privacy is to refer solely to privacy of digital identity, or sometimes the combination of that with privacy of personal communications. Privacy's importance is reflected in the fact that fundamental documents that define human rights all include reference to privacy or related ideas, such as the Universal Declaration of Human Rights [5] (UDHR, Article 12) states that "no one shall be subjected to arbitrary interference with his privacy, family, home or correspondence, nor to attacks upon his honor and reputation. Everyone has the right to the protection of the law against such interference or attacks" and in the International Covenant on Civil and Political Rights [6] (ICCPR 1966, Article 17) is expressed in very similar terms as "no one shall be subjected to arbitrary or unlawful interference with his privacy, family, home or correspondence, nor to unlawful attacks on his honor and reputation". In the 1950s European Convention on Human Rights [7], Article 8 is entitled right to respect for private and family life, and states that "everyone has the right to respect for his private and family life, his home and his correspondence. There shall be no interference by a public authority with the exercise of this right except such as is in accordance with the law and is necessary in a democratic society in the interests of national security, public safety or the economic well-being of the country,

G. Ben Ayed, *Architecting User-Centric Privacy-as-a-Set-of-Services*, Springer Theses,
DOI: 10.1007/978-3-319-08231-8_4,

for the prevention of disorder or crime, for the protection of health or morals, or for the protection of the rights and freedoms of others". The 2000s Charter of Fundamental Rights of the European Union [8] deals with privacy in Articles 7 and 8, and there are many specific European Directives. However, many national Constitutions and Bills of Rights also encompass privacy. However, Clarke [3] points out that governments and enterprises implement data protection and privacy laws to have a visibility when complying to international policies such as ICCPR rather than to respond to human rights needs.

4.1 Privacy: Preliminaries

Privacy is becoming an increasingly important field of research with many definitions and terminologies that are presented in the literature. In the UNESCO report [9], authors mention that the treasure of multiple privacy definitions is a consequence of multiple societal views of privacy and different privacy policies and regulations that are set with different intents, purpose, and outcomes. Merriam-Webster dictionary subdivide privacy into three elements, which are described broadly as follows: (1) the quality of state of being apart from the company; (2) the isolation, seclusion or freedom from unauthorized oversight or observation; or (3) a place of seclusion or retreat. Another privacy subdivision is suggested by two computer scientists at the University of Southampton: "being able to make your own decisions and hold your own views without interference; controlling information about yourself; and being in charge of your personal space, these basic elements of privacy are under threat" [10]. However, Clarke [3] restricts the scope of privacy to personal data protection and defines it as "the interest that individuals have in sustaining a personal space free from interference by other people and organizations". In fact, the notion of data protection derives from the 'fair information practices' movement that has been used by corporations and governments since the late 1960s to avoid meaningful regulation [2, 3]. Additionally, the same author mentions that privacy encloses four dimensions: (1) privacy of the person is concerned with the integrity of the individual's body, and is related to the Physiological and Safety levels of the Maslow's hierarchy such as compulsory immunization, imposed treatments such as lobotomy and sterilization, blood transfusion without consent, compulsory provision of samples of body fluids and body tissue, and requirements for submission to biometric measurement; (2) privacy of the personal behavior is related to both the Belonging and Self-Esteem levels of Maslow's hierarchy, and perhaps to Self-Actualization as well. Many issues that come to attention relate to sensitive matters, such as sexual preferences and habits, political activities and religious practices; (3) privacy of personal communications is referred to as 'interception privacy', is also related to both the Belonging and Self-Esteem levels of Maslow's hierarchy, and perhaps to Self-Actualization as well. Individuals desire the freedom to communicate among themselves, using various media, without routine monitoring of their communications by other persons or organizations. Issues include the directional microphones use with or without recording apparatus, the telephonic interception and recording, and the third-party access to email-messages, etc.; and (4) privacy of personal data is sometimes referred to as data privacy and information privacy, is again related to the upper layers of Maslow's hierarchy. Individuals claim that data about themselves should not be automatically available to other individuals and organizations, and that, even where data is possessed by another party, the individual must be able to exercise a substantial degree of control over that data and its use [3, 11].

Clarke [2, 3] thinks that privacy can be seen from a number of different perspectives. From the philosophical perspective, privacy is regarded being part of 'human dignity' and integrity that play a significant role in many countries and these ideas underpin the notion and significance of human rights. Psychologically, people need private space in the public arena as well as behind closed doors. Looking at privacy in sociological side, people need to be free to behave but without the continual threat of being observed. In economy, people need to be free to innovate. Innovators perceive themselves to be at risk, if they lack private space in which to experiment. Finally, politically, people need to be free to think, and argue, and act. In addition the author mentions that privacy-invasions are seriously harmful to the societies, economies and polities.

Privacy is becoming more important need and a research topic of keen interest in the era of digital and ubiquity. Computers and networking technologies have emerged and harnessed to the task of assisting governments and corporations to monitor people. Since, discussions about privacy protection have been largely focused on and limited to the protections of personal data, instead of people's interests. Within the same approach of data protection and specifically when they are disclosed to other parties, privacy is described as "the ability to determine for ourselves when, how, and to what extent information about us is communicated to others" [12]. In the offline world, privacy management issues arise from the blurring boundaries between the public and private spheres of the individual existence. However, in the online world, data collection is crossing the boundaries of space and time, with data about humans starting from pre-natal diagnostics to retirement daily life. Additionally, ubiquity is creating new opportunities for crossing more borders: natural borders, social borders, spatial borders, and temporal borders [13]. The advent of the digital leads to an increase in the amount, quality and accuracy of data generated and collected. The increase is not limited to data collection, but extends to data storing, analysis and process [14]. Moreover, the digital world provides a universal availability of data, an ease of its accessibility, its durability over time, and a possibility of its early and infinite accumulation. Many legal authorities point that data pertaining to the individual can be propagated only through the consent of the concerned individuals. Thus, many public and private organizations show a true awareness of privacy by making disclaimers when they acquire data. Currently, privacy concerns are being fueled by an ever increasing list of privacy violations, ranging from privacy accidents to illegal actions. Solove [15] presents a taxonomy of sixteen types of privacy violations that are classified into four categories: (1) information collection: surveillance and interrogation. When users are able to keep transaction contents confidential and to act anonymously, they protect themselves against surveillance threats. Systems that provide plausible deniability make it impossible for adversary to prove that the user is concealing information; (2) information processing: aggregation, identification, insecurity, secondary use and exclusion. The property that prevents the aggregation of information as related to each other or to a particular subject is unlinkability. Identification is connecting data to individuals. Anonymity, unlinkability and confidentiality properties prevent this connection to be revealed; (3) information dissemination: breach of confidentiality, disclosure, exposure, increased accessibility, blackmail, appropriation and distortion; (4) invasion: intrusion and decisional interference.

4.2 Digital Identity Management and Privacy

We present in this section, privacy properties definitions that have been subject of research. A consolidated proposal for terminology [16] and PRIME glossary [17] related to privacy and identity management propose a definition of privacy terminology and concepts related to digital identity and relationships between them. Anonymity is defined in the context of anonymity set from both sender and recipient perspectives as "a state of being not identifiable within a subjects' set, which is called the anonymity set". The anonymity set is the set of all possible subjects. Therefore, a sender may be anonymous only within a set of potential senders, his/her sender anonymity set, which itself may be a subset of all subjects worldwide who may send messages. The same can applicable to the recipient. Both anonymity sets may be disjoint, be the same, or they may overlap. In addition, the anonymity sets may vary over time. From the attacker's perspective, anonymity means that the attacker cannot sufficiently identify the subject within a set of subjects, the anonymity set. Unlinkability of two or more Items of Interest (IOIs, e.g., subjects, messages, actions, etc.) from an attacker's perspective means that "within the system, the attacker cannot sufficiently distinguish whether these IOIs are related or not". Undetectability of an item of interest (IOI) from an attacker's perspective means that "the attacker cannot sufficiently distinguish whether it exists or not". Unobservability of an item of interest (IOI) means "undetectability of the IOI against all subjects uninvolved in it and anonymity of the subject(s) involved in the IOI even against the other subject(s) involved in that IOI". There is a strong bound between identity and privacy, so in the context of digital identity management, privacy is defined as "the protection of the attributes, preferences and traits associated with an identity from being disseminated beyond the subject's needs in any particular transaction" [18]. Here is a case [19] to illustrate how privacy could contribute to the protection of attributes and preferences associated with Alex's digital identity. Alex wants to buy Vodka cooler for a college party, thus, he is required to produce proof of age to purchase the alcohol. He is not required to disclose data such as the name of his college or the address of his employer. Moreover, as Alex did pay in cash, neither his name, age nor license number were recorded. As such, Alex's early predilection for vodka will not be automatically communicated to his biology professor or to his parents. The privacy of his actions in this case is assured because the data in question is: (a) minimal: only a driver's license was presented, (b) temporary: the license was only examined briefly by the store clerk, and (c) un-linkable: it cannot be linked with Alex's other attributes (parents' name and address or professor's contact details) [19].

There is a strong relationship between digital identity, security and privacy. The 2006 ITU report [19] states that "digital identities are becoming an increasingly valuable commodity, and as a consequence, its protection and management has become a pressing matter". They allow subjects accessing online services and for this reason protecting and securing digital identities is a current major need and one of the major online business enablers. The author [18] provides more details about the nature of such relationship and demonstrates a circular relationship between digital identity, information security, and privacy. He explains that privacy is built upon a foundation of good information security, which is dependent on a good digital identity infrastructure. Authors [20] stress on the relationship between privacy and digital identity security. They explain that privacy management tools would play a key role to protect digital identity through accountability enforcement. Others such as [21] take technical approach and explains that privacy policies complaisance is one of the major requirements that need to be addressed as a part of any identity federation. The authors [21] apprehend trust, privacy, and attributes security as fundamental objectives, on which digital identity management systems should serve. Moreover, enforcing privacy would provide subjects a mean to control digital identities, which means giving subjects power over digital identities by showing them what attributes are on the web, how they can be exploited and what steps to take for reducing the risk of becoming a victim of digital identity fraud [21]. The relationship between digital identity management and privacy is demonstrated through consequences of how a bad management of digital identity harms privacy and puts enterprises at a risk. A government agency could risk damage through a leak of citizen's private information; a financial institution might incur financial penalties or brand degradation due to an unauthorized trade or withdrawal; a health care firm might suffer damaging lawsuits with the release of personal health information to the wrong parties; and finally a breach of security might put regulated organizations out of compliance with various related data privacy regulations and thus put them at risk of government enforcement actions [21]. Finally, user-centered DigIdM tools would provide better experience in dealing with privacy and confidential data management [20].

4.3 Digital Identity and Privacy Issues

When privacy is compromised, security of the individual, the organization or the country could be threatened. Privacy is considered as an integral part of digital identity management and it becomes more complex issue since digital technologies are breaking the borders of private and public spheres [22]. Below, we present few major issues related to identity and we do not intend to cover all of them.

4.3.1 Digital Identity Attributes Disclosure

Subjects are increasingly required to disclose more information about themselves and authenticate with their identities more often, thus, their privacy may be at greater risk [23]. Another privacy infringements are witnessed when Facebook or Google suddenly change the privacy settings causing members to reveal personal information unwittingly [24]. Gathering a lot of identity information could harm individual privacy. Although in many cases "less identification means more privacy" but in this case the opposite is true. Another example, when we aggregate shared identity information such as full, name, telephone number, and address may constitute a greater amount of information then a person fingerprint or DNA profile but the latter reveal much more about the individual. So small pieces of information that are shared with a multitude of parties may put the privacy at a higher risk than a larger amount of identity information that are accessed by authorized and trusted parties [25]. Preserving privacy could contribute to prevent from identity theft and avoid damages related to it such as unauthorized access, frauds, identity/profile data theft, harmed reputation, unfulfilled potential revenues, loss of potential customers, money laundering, impersonate business employees, cyber-crimes, and cyber terrorism.

4.3.2 Digital Identity Attributes Processing and Analysis

The gathering, processing and analysis of information are crucial aspects of today's digital information economy. Without it, cash would be required for every purchase; there would be no licensed drivers, no health system, and no unemployment benefits. To create, use, store, and verify identity in the Internet is a complex issue that impacts society and individuals (example: privacy), corporations (corporate regulation), and governments (law, regulation, international treaties) [18]. Many public and private organizations show a true awareness of privacy by making disclaimers when they acquire data [19]. The gathering, processing and analysis of information are crucial aspects of today's digital information economy. Defining the limits of data collection relating to human individuals and the safeguarding of authorized data are matters of too great an importance. Thus, a delicate balance between the need to harness acquired and accumulated data for economic progress, quality of life and convenience; and the need to maintain privacy. Biometric data is now being used in many cases for identification purposes, or for entry into a particular country, notably in the United States through its US-VISIT program, under which foreign visitors are required to provide fingerprints upon entry [15, 19, 26–28]. Based on the study [29], a Single Identification Number (SIN) is under consideration in a number of EU countries (78 % in 2005) but the, the application of the SIN is not harmonized neither at the global nor European scale. In terms of the number of data linked to SIN, there is no consensus among EU countries. Some countries limit the data to those items that are absolutely necessary (less than ten in France, Italy and Lithuania) but many others, however, have identified a wider array of data (over twenty five in Bulgaria and Cyprus). Currently in EC, a number of countries are conducting debates about the number of data attributes needed, the legislative and organizational framework to regulate the use of the SINs, and the role of the designated supervisory authority. Privacy violations are taking place without the knowledge of consumers, and in some cases, consumers are left with little choice if they are to adopt new services. From privacy and security perspectives, an environment in which citizens are obliged to disclose more and more personal data, simply in exchange for convenience, or for lower prices, must be discouraged and eventually eliminated. For example, on the internet today, most are obliged (usually by default) to accept cookies that track online behavior—a phenomenon that just a few years ago was considered to be a serious invasion of privacy. Another common issue is that users' data can easily change hands thus shifting contractual obligations such as in Google case. When the company purchased Usenet in 2001, it acquired all the personal data that Usenet had collected. Google gave no guarantees about removing those data from its repositories. The same issue arose following the sale of eGroups to Yahoo [26].

4.3.3 Digital Identity Persistence and Visibility

A coincidence between happening related to a subject and storage is resulting privacy issues. In criminal cases, psychological profiling has given way to DNA matching. In consumer products, commodity logistics have given way to RFID databases. Genomics are the universal identification of life in the abstract; biometrics is considered as the universal identification of life in the particular; collaborative filters are the universal identification of life in the relational [30]. Offline activities of many people are tracked by CCTV cameras, Oyster cards and RFID tags, the details of the online searches and purchases accumulate in databases and many people also broadcast their lives through Web 2.0 sites such as blogs and social networking. Privacy is not the same old notion when it deals with digital self [10]. The same authors add that the attitude of people towards privacy may be originated from a lack of understanding the fact that in the online world the memory of an action will outlast and the audience is much wider than your close relatives and friends [10]. As mentioned in Sect. 3.6, digital identity attributes persistence issue has been a result of digitization. Therefore, privacy and attributes control mechanisms and polices should turn attributes from long-term memory into short-term memory by limiting and discouraging the recall process and promoting a 'delete' capability that would allow to move from unforgettable and unforgivable to forgettable and forgivable network and society [31]. The sociologist Armand Mattelart [30] points out that the current century is the era of universal standards of identification and local-ability. In this context, RFID tag enables to uniquely identifying a subject in extended business models. However, RFID usage is promoting a power's shift towards the manufacturer, who may gain the ability to track products across the supply chain independently of retailers. Therefore, privacy concerns over the insecurity of data exchange mechanism and the lack of identity-masking capability for current RFID technology are leading many privacy activists to oppose the embedding of RFID tags in consumer products, official documents, and so on. Many consumers are still uncomfortable with the idea that the can of beans that they bought could be tracked anywhere, by anyone with access to an RFID reader [32, 33]. From attributes security and control perspective, we proposed in chapter three an approach to weak the link between attributes, in a form of linked digital identity documents, to make them less visible. This could contribute to weak the digital identity persistence. Google provides a multi-languages Dashboard utility that allows users to display information associated with their Google Account. Once logged-in, Dashbord[1] page offers a view of several personal information related to a particular service. This is part of Google's efforts to provide to users more control over their personal information [34, 35]. However, if a user deletes his web navigation history, is it also deleted from Google servers? Until now, we are incapable to do so but we could make digital identity less visible. Within the same perspective, concepts of 'trash' and 'nonexistence' [30] are suggested. The trash always contains traces and signatures of use such as monthly bills, receipts, personal papers, cellophane wrapping, price tags, and spoiled food. Putting identity into a trash means promoting less visibility in order to give to the subject less control loss over digital identity. The same author highlights that future avant-garde practices will be those of nonexistence. A simple laser pointer can blind a surveillance camera when the beam is directed at the lens. In consequence, one is not hiding, simply nonexistent to that node [30]. Visibility of digital identity in social networks could be consequences of risky behavior such as leaving privacy settings as default 'open'; exhibiting sensitive personal information and exposing private life; and contacting or accepting unknown 'friends' people [36, 37].

Web companies should agree on a common mechanism that allows users to keep their information from being searched, as a means to preserve privacy and secure identity. In addition, legislation requiring opt-out controls may be needed [38]. The author [39] mentions that Internet search companies, such as Google [40–43], publicly say that they protect users' privacy by encrypting personal information and by using numbers instead of names to give their users anonymity. However, anonymization is not always effective. He provides a case to demonstrate that treatment of anonymous personal information could reveal user identity. He says: "AOL user

---

[1] http://www.google.com/dashboard

number 4417749 found this out the hard way in 2006 when AOL decided to publish online a list of 20 million Web searches, including hers and those of 657,000 other users. Reporters were able to track down the 62-year old widow in Lilburn, Ga., by analyzing the content of her searches. Luckily, Thelma Arnold was relatively unembarrassed by the revelation of her identity and intimate interests" [39]. Moreover, the same author adds that adopting a new quantum version of the Web in which communication is ensured via quantum encryption would enable the user to send queries and receive answers with the assurance that no one—not even Google—knows what questions you have asked. With such technology private searching will be guaranteed during the online experiences. However, the need of quantum Web search would definitely push search engine companies to reconsider their business models. Currently, search engines save and analyze users' data and behavior to be able to display targeted ads in order to make a profit. In the near future, with private search, search engines will need a new business model and users may have to pay for their search since quantum communication is still expensive [39]. Companies could also provide options at the choice of the user such as free Web search and charged quantum Web search, which could be included in the bill of Internet connection as the same as the international calls in phone bills.

4.3.4 Loosely Coupled Collaborative IS, Digital Identity and Privacy

Recent years have seen the trend of business globalization which urgently requires dynamical collaboration among organizations. The business processes of different organizations need to be integrated seamlessly to adapt the continuously changing business conditions and to stay competitive in the global market. Though current business process technologies have achieved a certain level, there is still a large room between the current supports and the requirements from real collaboration scenarios. Especially in a loosely coupled collaboration environment, many non-functional yet crucial aspects, such as privacy and security, are with a great lack of sufficient supports. Collaborative environments present major challenges to privacy since collaboration involves the exchange of digital identities between collaborators [44]. There is a need to establish a balance between the benefits of collaborative environments, which provide knowledge discovery and sharing against the protection of individual and organizational privacy needs [45].

4.4 Privacy Policies

We cover in this section privacy policies and regulations that are related to digital identity. They are classified into three main categories: global, domestic, and business-specific privacy policies.

4.4.1 Global Privacy Policies

Global privacy policies related to digital identity are principles, frameworks, and guidelines that are suggested by international bodies, regional policy-makers and global legal framework to provide common practices in order to help organizations to make global-scale business. In addition, it encompasses also the requirements that are neither domestic nor business-specific but practices and assessment tools such as Fair Information Practices and Privacy Impacts Assessment Tools that are provided by organizations having a global vision.

4.4.1.1 CDT's 2007 Privacy Principles for Identity in the Digital Age

In a draft for comment [25], the center of democracy and technology, a non-profit public interest organization that works to enhance free expression and privacy in communications technologies, proposed an identity-related privacy framework that comprises eleven 'privacy principles' to encourage public and private sectors entities to develop systems involving the collection, authentication, and use of identity information. The privacy principles are classified into two types; the first three are overarching principles that are particularly relevant to digital identity and the rest are FIPs-based principles, which are adaptation of the widely recognized FIPs to the identity context. Each of the principles is explained as follows: (1) diversity and decentralization: using a centralized identity solution or a single identifier or credential for multiple purposes diminishes the ability of identity system to protect privacy. If this deemed necessary, strong safeguards should be addressed in the design phase to ensure that unnecessary linkages do not occur; (2) proportionality: identity system can collect larger amounts and/or more sensitive identity information, such as race, ethnicity, and religious and political affiliation, from individuals seeking to participate in transaction of higher significance; (3) privacy and security by design: privacy consideration should be incorporated into an identity system from the outset of the design process; (4) purpose specification: the purposes of the identity system and for which identity-related information will be collected and used should be clearly defined; (5) limited use: identity, authentication, and linked information should be used and retained only for the specific purposes for which they were collected; (6) notice: individuals should be provided with clear statement about the collection and use of identity, authentication, and linked information; (7) individual control and choice : an identity system should offer individuals reasonable, granular control and choice over the attributes and identifiers needed to enroll in the system and the credentials that can subsequently be used within the system; (8) security: organizations that handle identity, authentication, and linked information should provide reasonable technical, physical, and administrative safeguards to protect against loss or misuse of the information; (9) accountability: organizations that handle identity, authentication, and linked information should be able to verify that they are complying with applicable privacy and security protections; (10) access: individuals should be provided reasonable access to the identity, authentication, and linked information that organizations maintain about them and use in the ordinary course of business; (11) data quality: organizations should strive to ensure that the identity information they hold is timely, complete, and accurate.

4.4.1.2 OECD's 1980 Guidelines on the Protection of Privacy and Transborder Flows of Personal Data

OCDE [46] provides eight basic principles of the protection of privacy and transborder flows of personal data: (1) collection limitation: collecting data about subjects has to be within fairness, through lawful means, and with the knowledge or consent of the subject; (2) data quality: data should be relevant and appropriate for the purpose of data collection and usage; (3) purposes specification is a prerequisite of data collection and should be set before the time of the collection. Moreover, the subsequent use of data is limited only to fulfill the pre-specified purpose; (4) use limitation: subject's data are to be disclosed, made available or used only to fulfill the specified purpose with the subject's consent or by the authority of law; (5) security safeguards protect subject's data against risks such as loss, unauthorized access, destruction, use, modification or disclosure of data; (6) openness about developments, practices and policies with respect to personal data; (7) subject participation: the subject should have the right to obtain from data controller a confirmation of whether or not he has data relating to him. Data relating to subject is to be communicated within reasonable time and in reasonable manner, at not excessive charges if any, and in a form that is readily intelligible to him. In addition, the subject has the right to clearly receive reasons if the request that is made under subparagraphs is denied and to challenge such a denial. Finally, he can challenge data relating to him and he would have the data erased, rectified, completed or amended; (8) accountability: a data controller should be accountable for complying with measures which give effect to the principles stated above.

4.4.1.3 OECD's 2008 Data Protection and User Control for Identity Management Systems

OCDE report [47] mentions that the demand of identity management tools is likely to increase if they allow: (1) notice of other parties' treatment of identity information, (2) an opportunity for the user to consent to or refuse this treatment; (3) an assurance of security in which privacy is highlighted. The report presents seven aspects of privacy that are bounded to user control: (a) decentralization of identity data into a maximum of separate data contexts and stakeholders; (b) data minimization means minimum of identity data that are necessary to support all the required transactions, should be stored; (c) local identifier is needed because each context should whenever possible using local pseudonyms to identify the set of identity data associated with the person. From this perspective, global/universal identifier setup is discouraged; (d) verifiability of the user's claim by the relying party. The identity system should support a verification mechanism of the claims. In other words, relaying parties require that the claim made about the user be verifiable; (e) selective disclosure, beyond the minimum of data to be stored, only identity information that are needed for a specific transaction should be involved; (f) composability provides to the user the ability to aggregate reusable groups of related partial identities into a convenient digital profile that can be used in recurring needs such as commercial transactions. Without this possibility, the user would rely on smaller, less-minimal and easy correlated digital identities that will reduce his privacy; (g) auditability of the identity infrastructure means allowing audit in order to provide accountability and enable records to be legally redressed; and (4) access to information on actual practices affecting their data, with an opportunity for redress.

4.4.1.4 (95/46/EC1) European Union Data Protection Directive

The Data Protection Directive (officially Directive 95/46/EC) [48] regulates the processing of personal data within the European Union. It is considered as an important component of EU privacy and human rights law. All the member states of the European Union (EU) are signatories of the European Convention on Human Rights (ECHR). Article 8 of the ECHR provides a right to respect for one's "private and family life, his home and his correspondence" subject to certain restrictions. The directive enunciates that digital identity attributes should not be processed at all, except when certain conditions are met. These conditions fall into three categories: (1) transparency: subject has the right to be informed when his attributes are being processed. The responsibility for compliance rests on the shoulders of the controller, who must provide his name and address, the purpose of processing, the recipients of the attributes and all other information required to ensure the processing is fair (articles 10 and 11). Attributes may be processed only under the following circumstances (article 7): (a) when the subject has given his consent; (b) when the processing is necessary for the performance of or the entering into a contract; (c) when processing is necessary for compliance with a legal obligation; (d) when processing is necessary in order to protect the vital interests of the subject; (e) processing is necessary for the performance of a task carried out in the public interest or in the exercise of official authority vested in the controller or in a third party to whom attributes are disclosed; (f) processing is necessary for the purposes of the legitimate interests pursued by the controller or by the third party or parties to whom attributes are disclosed, except where such interests are overridden by the interests for fundamental rights and freedoms of the subject. The subject has the right to access all attributes processed about him. The subject even has the right to demand the rectification, deletion or blocking of data that is incomplete, inaccurate or isn't being processed in compliance with the data protection rules (article 12); (2) legitimate purpose: attributes can only be processed for specified explicit and legitimate purposes and may not be processed further in a way incompatible with those purposes (article 6b); and (3) proportionality: attributes may be processed only insofar as it is adequate, relevant and not excessive in relation to the purposes for which they are collected and/or further processed. Attributes must be accurate and, where necessary, kept up to date; every reasonable step must be taken to ensure that data which are inaccurate or incomplete, having regard to the purposes for which they were collected or for which they are further processed, are erased or rectified. Attributes shouldn't be kept in a form which permits identification of subjects for longer than is necessary for the purposes for which the attributes were collected or for which they are further processed. EU members shall lay down appropriate safeguards for digital identities stored for longer periods for historical, statistical or scientific use (article 6). When sensitive attributes such as religious beliefs, political opinions, health, sexual orientation, race, and membership of past organizations are being processed, extra restrictions apply (article 8). Subject may object at any time to the processing of attributes for the purpose of direct marketing (article 14). A decision which produces legal effects or significantly affects the subject may not be based solely on automated processing of attributes (article 15). A form of appeal should be provided when automatic decision making processes are used [49].

4.4.2 Domestic Privacy Policies

We cover in this section different privacy acts and policies related to digital identity that are presented by national bodies and local privacy authorities in United States, Canada, Japan, and Australia.

4.4.2.1 The United States Privacy Act of 1974

Privacy is embodied, not stated, in the US. Bill of Rights [50]. However, the United States Privacy Act of 1974, which has been in effect since 1975, attempts to regulate the collection, maintenance, use, and dissemination of digital identity attributes by federal agencies. The Act requires the agencies to (1) permit an individual to determine what records pertaining to him are collected, maintained, used, or disseminated; (2) permit an individual to prevent records pertaining to him obtained for a particular purpose from being used or made available for another purpose without his consent; (3) permit an individual to gain access to information pertaining to him in records, and to correct or amend such records; (4) collect, maintain, use or disseminate any record of personally identifiable information in a manner that assures that such action is for a necessary and lawful purpose, that the information is current and accurate for its intended use, and that adequate safeguards are provided to prevent misuse of such information; (5) permit exemptions from the requirements with respect to the records provided in this Act only in those cases where there is an important public policy need for such exemption as has been determined by specific statutory authority; and (6) be subject to civil suit for any damages which occur as a result of willful or intentional action which violates any individual's right under this Act [51, 52]. Moreover, U.S. Privacy Act of 1974 requires that any federal, state, or local government agency that requests your Social Security Number (SSN) must tell you four things: (1) whether disclosure of your SSN is required or optional; (2) what statute or other authority requires this number; (3) how they will use your SSN, once they have it; and (4) what will happen if you do not provide them with your SSN [53].

4.4.2.2 CSA Model Code for the Protection of Personal Information of 1996

Committed to the protection of privacy, the Canadian government signed in 1984 the Organization for Economic Co-operation and Development (OECD) Guidelines on the Protection of Privacy and Transborder Flows of Personal Data. The OECD Guidelines were used as the basis for the development of CSA Model Code for the Protection of Personal Information [54]. The CSA Model Code is similar to the OECD guidelines. The major differences are that the CSA Model Code makes consent and disclosure limitation separate principles, and adds retention limitation as a new principle [69]. The Standard addresses two broad issues: the way organizations collect, use, disclose, and protect personal information; and the right of individuals to have access to personal information about themselves, and, if necessary, to have the information corrected. Ten interrelated principles form the basis of the Standard: (1) Accountability; (2) identifying purposes; (3) consent; (4) limiting collection; (5) limiting use, disclosure, and retention; (6) accuracy; (7) safeguards; (8) openness; (9) individual access; and (10) challenging compliance [54].

4.4.2.3 The Canadian Personal Information Protection and Electronic Document Act of 2000

In Canada, the key elements of the Privacy Code are now incorporated into the Personal Information Protection and Electronic Documents Act (PIPEDA) [55]. All organizations that comply with the CSA standard, are meeting the federal requirements of PIPEDA [56].

4.4.2.4 The Canadian Privacy Act of 1983

The Privacy Act [57] is Canadian federal legislation that came into effect on July 1st, 1983. The act sets out rules for how institutions of the federal government must deal with personal information of individuals. Some salient provisions of the legislation are as follows: (1) a government institution may not collect attributes unless it relates directly to an operating program or activity of the institution (section 4); (2) with some exceptions, when a government institution collects an digital identity attributes from the subject, it must inform the individual of the purpose for which the information is being collected [section 5(2)]; (3) with some exceptions, digital identity attributes under the control of a government institution may be used only for the purpose for which the information was obtained or for a use consistent with that purpose, unless the individual consents (section 7); (4) with some exceptions, attributes under the control of a government institution may not be disclosed, unless the individual consents (section 8); (5) every Canadian citizen or permanent resident has the right to be given access to subject's digital identity under the control of a government institution that is reasonably retrievable by the government institution, and request correction if the information is inaccurate (section 12); (6) the privacy commissioner of Canada receives and investigates complaints, including complaints that an individual was denied access to his or her attributes held by a government institution (section 29).

4.4.2.5 The Japanese Act on the Protection of Personal Information of 2003

Japan enacted the Personal Information Protection Act (JPIPA) [58] in 2003 to protect individuals' rights and personal information while preserving the benefits of information technology and personal information. The law establishes responsibilities for businesses that handle citizens' attributes of Japan and outlines potential fines and punishments for organizations that do not comply. The act requires organizations to communicate their purpose in collecting and using personal information. They must also protect personal information from disclosure, unauthorized use or destruction [59]. According to the act, a subject that handles attributes shall: (1) specify the purpose for collecting and using personal information; (2) not acquire information by fraudulent or other unfair means; (3) ensure that personal data are kept secure from loss and unauthorized access and disclosure; (4) promptly notify the subject of the purpose for which his attributes will be used, or otherwise announce the purpose for use; (5) refrain from supplying personal data to third parties without the prior consent of the individual concerned, except in certain defined circumstances (presumably the restrictions on providing information to third parties covers transfers both inside and outside the country); (6) respond to the subject requests for correction, supplementation or deletion of personal data; (7) respond to the subject requests that an entity cease using personal information altogether; (8) endeavor to appropriately and promptly handle individual complains about the handling of attributes. Businesses must also endeavor to set up an internal complaint-handling system [60].

4.4.2.6 The Australian Privacy Act of 1998 (Private Sector)

The IPPs regulate how Australian and ACT government agencies manage personal information. The major objectives are how and when personal information can be collected, how it should be used and disclosed, and storage and security. Moreover, the principles allow individuals to access personal information and correct it if it is wrong. Here is a plain English summary of the eleven Information Privacy Principles (IPPs): (1) manner and purpose of collection: the information must be necessary for the agency's work, and collected fairly and lawfully; (2) collecting information directly from individuals: an agency must take steps to tell individuals why they are collecting personal information, what laws give them authority to collect it, and to whom they usually disclose it; (3) collecting information generally: an agency must take steps to ensure the personal information it collects is relevant, up-to-date and complete and not collected in an unreasonably intrusive way; (4) storage and security: personal information must be stored securely to prevent its loss or misuse; (5–7) access and amendment : these principles require agencies to record the type of personal information that they hold and to give individuals access to personal information about them. Personal information can be amended or corrected if it is wrong; (8–10) information use: these principles outline the rules about keeping accurate, complete and up-to-date personal information; using information for a relevant purpose; and only using the information for another purpose in special circumstances, such as for some health and safety or law enforcement reasons and of course with the individual's full consent; (11) disclosure: this principle sets out when an agency may disclose personal information to someone else, for example another agency. This can only be done in special circumstances [61].

4.4.2.7 The Swiss Federal Law on Personal Data Protection (1992)

The law specifies that the collection of personal data cannot be done only in a lawful manner, whether in the form of consent of the person concerned, of a public or private interest, or law. Under the principle of proportionality, only the necessary data that enable to meet the target can be treated. The principle of finality is that the data collected are treated only to the extent necessary to achieve the goal on which parties have agreed on during their collection. The right of access to data for the individual concerned should allow him to assert his rights, in particular by requesting the correction or deletion of data concerning him. In accordance to the article 8, the person has a right to know whether information is processed on or disclosed to third parties [62].

4.4.2.8 The French Data Protection and Freedoms Act

The Act no. 78-17 of 6 January 1978 on Data Processing, Data Files and Individual Liberties, hereafter referred as the French Data Protection and Freedoms Act (DPA), imposes to organizations that implement data processing or hold data files must guarantee their security. Personal data should be collected and processed in an fair and lawful manner and collected for determined, explicit and legitimate purposes and is not later on processed in a way that is incompatible with these purposes (article 6). Personal data should also be preserved in a form allowing the identification of the persons concerned for a period of time which shall not exceed the duration required by the purposes for which it is collected and processed. A mechanism for suppression, archiving, or anonymization of this data when its retention period expires should be available. A risk management represents an effective way to protect the fundamental rights and freedoms of natural persons and in particular their right to privacy with respect to the processing of personal data (article 1 of Directive 95/46/EC) [63].

4.4.2.9 The Law of Personal Data Protection in Tunisia (2004)

The law num 2004-63 July 27, 2004, relating to the protection of personal data in Tunisia (article 9) stipulates the processing of personal data must be made within the framework of respect for human dignity, privacy and civil liberties. It prohibited to use personal data or processed personal data to harm the people or their reputation. The goal of each personal data processing operations should be clearly stated with the consent of the persons and authorization of the national authority of personal data protection. Decree No. 2007-3003 of 27 November 2007 laying down the operating procedures of the national authority of personal data protection [64].

4.4.3 Business-Specific Privacy Policies

The business-specific requirements represent an industry or domain-specific requirement such as health, finance, education, and transportation sectors.

4.4.3.1 The 1996 Health Insurance Portability and Accountability Act

The act provides to patients control over how their medical records are used and disclosed [65]. The HIPAA Privacy Rule regulates the use and disclosure of subject's Protected Health Information (PHI) held by 'covered entities' (generally, health care clearinghouses, employer sponsored health plans, health insurers, and medical service providers that engage in certain transactions). PHI is any information held by a covered entity which concerns health status, provision of health care, or payment for health care that can be linked to an individual. This is interpreted rather broadly and includes any part of subject's medical record or payment history. Covered entities must disclose PHI to the individual within 30 days upon request. A covered entity may disclose PHI to facilitate treatment, payment, or health care operations, or if the covered entity has obtained authorization from the subject. However, when a covered entity discloses any PHI, it must make a reasonable effort to disclose only the minimum necessary information required to achieve its purpose. Subjects have the right to request that a covered entity correct any inaccurate PHI. It also requires covered entities to take reasonable steps to ensure the confidentiality of communications with subjects. For example, a subject can ask to be called at his or her work number, instead of home or cell phone number. Covered entities have to notify subjects of uses of their PHI. Covered entities must also keep track of disclosures of PHI and document privacy policies and procedures. They must appoint a Privacy Official and a contact person responsible for receiving complaints and train all members of their workforce in procedures regarding PHI. Any subject, who believes that the Privacy Rule is not being upheld, can file a complaint with the Department of Health and Human Services Office for Civil Rights [66].

4.4.3.2 The 1999: Bliley Financial Services Modernization Act

The Gramm–Leach–Bliley Act (GLB), also known as the Financial Services Modernization Act is enacted November 12th, 1999 to govern the collection, disclosure, and protection of consumers' nonpublic personal information; or personally identifiable information. Financial Privacy Rule requires financial institutions to provide each consumer with a privacy notice at the time the consumer relationship is established and annually thereafter. The privacy notice must explain the information collected about the consumer, where that information is shared, how that information is used, and how that information is protected. The notice must also identify the consumer's right to opt out of the information being shared with unaffiliated parties. Should the privacy policy change at any point in time, the consumer must be notified again for acceptance. Each time the privacy notice is reestablished, the consumer has the right to opt out again. The unaffiliated parties receiving the nonpublic information are held to the acceptance terms of the consumer under the original relationship agreement [67].

4.5 Digital Identity-Related Privacy Requirements

We draw DigIdeRP requirements [68] from policies of three types of initiatives regarding privacy: global, domestic, and business-specific privacy policies related to digital identity. DigIdM systems should be fully designed in accordance of the following DigIdeRP requirements.

4.5.1 Purpose Specification of Attributes Collection

Digital identity attributes that have been collected shall be associated with the purpose. In addition, the answer of the following question must be provided: why specific data are being collected? As a consequence, attributes processing or communication should be in a consistence with the purposes for which attributes has been collected. The purpose of the system and the purposes for which identity information will be collected and used should be directly linked. Each purpose should have a clear and publicly communicated rationale behind it. For instance, if data were collected for medical treatment purpose, thus querying data for drug marketing purpose will not be possible. The amount, sensitivity and type of identity information collected from subjects should be proportional, to the purpose for which it is collected. Sensitive identity information such as race, ethnicity, or religious or political affiliation, should be anonymized to the greatest extent possible.

4.5.2 Consent for Attributes Usage/Release

The subject provides his consent for usage of the attributes that they have provided for the specific purpose. For instance, a user can give consent for his attributes to be released for medical research purposes. Individuals should be notified when other information is gathered about them and linked to their identity.

4.5.3 Limited Usage of Attributes

Attributes that are collected shall be limited to the minimum necessary for accomplishing the specified purposes. For instance, requirement of bank account number for medical records is absurd. Identity, authentication, and linked information should be used, shared and retained only for the specific purposes for which they were collected/shared/retained.

4.5.4 Limited Retention of Attributes

Attributes shall be retained only for the necessary period of the purpose's fulfillment for which it has been collected. For instance, a patient medical history can only be retained for a period of 12 months after the treatment, unless the patient has given attributes release consent for research purpose.

4.5.5 Accuracy of Stored Attributes

Attributes that are stored in the database shall be accurate and up-to-date. For instance, administering a wrong medication to a patient due to outdated attributes in his medical record may cause serious injury and illness.

4.5.6 *Openness*

The subject should be able to access to his stored data. Attributes should be easy for subjects to access, view, understand and change. Subjects should also be able to challenge conclusions drawn from digital identity aggregation. Whenever possible, subjects should be able to see when their identity attributes has been disclosed and to whom.

4.5.7 *Authentication and Enrollment Needs*

Subject's enrollment and authentication should be with different identities for different purposes. Subjects should be allowed to choose the appropriate authentication means to satisfy a specific need within a single system. It is not optimal to centralize identity information or use a single credential for a multitude of purposes. Using a single identifier or credential for multiple purposes creates a single target for privacy and security abuses. When linking attributes within different systems is deemed necessary, appropriate safeguards should be implemented to limit the associated privacy and security risks.

4.5.8 *Choice and Terms of the Contract*

A system should offer individuals reasonable, granular control and choice over the attributes and identifiers needed to enroll in the system and the credentials that can subsequently be used within the system. Moreover, if an individual declines to accept the terms of contract, no information should be collected. When possible, individuals should be able to consent to participation in an identity system but decline particular terms of the contract.

4.5.9 *Secondary Use*

Secondary use, sharing, and sale of identifiers or credentials should not be permitted. Thus, multiple uses of identifiers and credentials should be avoided particularly in the authentication context. Identity, authentication and linked information should be shared with third parties including data transfers between government and commercial entities only when necessary, and should be stored by third parties only until the purpose for which it was shared has been completed.

4.5.10 Compliance

A subject should be able to check privacy compliance with the above principles. For instance, a patient should be able to see that privacy policies concerning his attributes are enforced. This would gain the trust of the patient.

4.5.11 Project-Specific Privacy Requirements

We let this requirement open to privacy needs that are not stated above and that would be articulated by the user for a project-specific purpose.

References

1. B. Schneier, The eternal value of privacy (2006), Available: http://www.wired.com/politics/security/commentary/securitymatters/2006/05/70886. Accessed 5 May 2010
2. R. Clarke, Introduction to dataveillance and information privacy, and definitions and terms (1999), Available: http://www.anu.edu.au/people/Roger.Clarke/DV/Intro.html. Accessed 6 May 2010
3. R. Clarke, What's Privacy? [Workshop at the Australian Law Reform Commission] (2006), Available: http://www.anu.edu.au/people/Roger.Clarke/DV/Privacy.html. Accessed 6 May 2010
4. M.H. Kamali, *The Dignity of Man: An Islamic Perspective*, 2nd edn. (Islamic Texts Society, 2002)
5. U. Nations, The universal declaration of human rights (1948), Available: http://www.un.org/en/documents/udhr/index.shtml. Accessed 1 April 2011
6. The Office of the United Nations High Commissioner for Human Rights, International Covenant on Civil and Political Rights (1966), Available: http://www2.ohchr.org/english/law/ccpr.htm. Accessed 1 April 2011
7. European Convention on Human Rights (1950), Available: http://conventions.coe.int/Treaty/en/Treaties/Html/005.htm. Accessed 1 April 2011
8. European Union, The charter of fundamental rights of the European Union (2000), Available: http://ec.europa.eu/justice_home/fsj/rights/charter/fsj_rights_charter_en.htm. Accessed 1 April 2011
9. G. Hosein, Politics of information society: the bordering and restraining of global data flows (2004), Available: http://www.privacyinternational.org/survey/censorship/unesco.pdf. Accessed 6 May 2010
10. K. O'Hara, N. Shadbolt, *The Spy in the Coffee Machine: The End of Privacy as We Know It* (Oneworld Publications, 2008)
11. L.S. Nelson, Constructing policy: the unsettled question of biometric technology and privacy, in *Privacy and Technologies of Identity: a Cross-disciplinary Conversation* (Springer, Berlin, 2006), pp. 151–172
12. Privacy, in *Stanford Encyclopedia of Philosophy* (2006)
13. G.T. Marx, Murky conceptual waters: the public and the private. Ethics Inf. Technol. **3**(3), 157–169 (2001)
14. J. Cas, Privacy in pervasive computing environments—a contradiction in terms? IEEE Technol. Soc. Mag. **24**(1), 24–33 (Spring, 2005)

15. D.J. Solove, A taxonomy of privacy. J. Univ. PA Law Rev. **154**, 477 (2006)
16. A. Pfitzmann, M. Hansen, Anonymity, unlinkability, unobservability, pseudonymity, and identity management—a consolidated proposal for terminology (2008), Available: http://dud.inf.tu-dresden.de/literatur/Anon_Terminology_v0.31.pdfIdentity. Accessed: 21 May 2010
17. PRIME, PRIME glossary related to privacy and identity management, Available: https://prime.inf.tu-dresden.de/prime/space/Dictionary. Accessed: 11 May 2008
18. P.J. Windley, *Digital Identity: Unmasking identity management architecture (IMA)* (O'Reilly Media, 2005)
19. International Telecommunication Union, Digital life. ITU internet report (2006), Available: http://www.itu.int/osg/spu/publications/digitalife/docs/digital-life-web.pdf. Accessed 21 May 2010
20. M.C. Mont et al., Towards accountable management of privacy and identity information. LNCS (Springer, Berlin, 2003)
21. M. Gardiner, The business value of identity federation (2007), Available: http://whitepaper.techworld.com/authentication/4818/the-business-value-of-identity-federation. Accessed: 21 May 2010
22. D. Boyd, Why youth (heart) social network sites: the role of networked publics in teenage social life. MacArthur Foundation Series on digital learning—youth, identity, and digital media volume (2007), Available: http://www.danah.org/papers/WhyYouthHeart.pdf. Accessed 27 Aug 2010
23. Organizing Committee of Digital Identity & Privacy (Human Capital & Social Innovation Technology Summit), Call for controbution to managing digital identities for education, employment and business development (2007), Available: http://events.eife-l.org/HCSIT2007/overview/dip/dip2007. Accessed 11 May 2010
24. K. Cukier, The data deluge: businesses, governments and society are only starting to tap its vast potential. The economist (February 23–March 5) (2010), Available: http://www.economist.com/opinion/displaystory.cfm?story_id=15579717. Accessed 13 May 2010
25. Center for Democracy & Technology, Privacy principles for identity in the digital age [draft for comment—version 1.4] (2007), Available: http://www.cdt.org/files/pdfs/20071201_IDPri vacyPrinciples.pdf. Accessed 28 May 2010
26. H. Noonan, Identity, in *Stanford Encyclopedia of Philosophy* (2009)
27. K.J. Strandburg, Social norms, self control, and privacy in the online world, in *Privacy and Technologies of Identity: a Cross-disciplinary Conversation* (Springer, Berlin, 2006), pp. 31–54
28. T.Z. Zarsky, Online privacy, tailoring, and persuasion, in *Privacy and Technologies of Identity: A Cross-disciplinary Conversation* (Springer, Berlin, 2006), pp. 209–224
29. B. Otjacques et al., Identity management and data sharing in the European Union, in *Proceedings of the 39th Hawaii International Conference on System Sciences*, 2006
30. A.R. Galloway, E. Thacker, *The Exploit—A Theory of Networks* (University of Minnesota Press, 2008)
31. V. Mayer-Schönberger, *Delete: The Virtue of Forgetting in the Digital Age* (Princeton University Press, Princeton, 2009)
32. J. Madelin, L. Razzell, Towards the Identity Society, Available: http://www.identitysociety.org/files/identitysociety.pdf. Accessed 7 Sept 2009
33. A. Juels, RFID privacy: a technical primer for the non-technical reader, in *Privacy and Technologies of Identity: a Cross-disciplinary Conversation* (Springer, Berlin, 2006), pp. 57–74
34. Google, Google privacy principles, Available: http://www.google.com/intl/en/corporate/privacy_principles.html. Accessed 14 May 2010
35. Google Privacy Center, Privacy FAQ. Available: http://www.google.com/intl/en/privacy_faq.html#serverlogs. Accessed 14 May 2010
36. Ofcom: Office of Communication, Social networking: a quantitative and qualitative research report into attitudes, behaviours and use (2008), Available: http://stakeholders.ofcom.org.uk/binaries/research/media-literacy/report1.pdf. Accessed 27 Aug 2010
37. N. Reichenthal, Une identité connectée. Bulletin HEC—Le Magazine des Gradués (2007), Available: http://www.gradueshec.ch/bulletins/documents/75nadine.pdf. Accessed 13 May 2010
38. S. Davies, Opt-Out control. Scientific American (2007)
39. S. Lloyd, Privacy and the Quantum Internet Scientific American Magazine (2009)

40. Google—Centre de confidentialité, Publicité et confidentialité (2010), Available: http://www.google.com/privacy_ads.html. Accessed 28 Aug 2010
41. Google—Centre de confidentialité, Règles de confidentialité (2009), Available: http://www.google.com/intl/fr/privacypolicy.html. Accessed 28 Aug 2010
42. Google—Centre de confidentialité, FAQ relative à la confidentialité (2010), Available: http://www.google.com/intl/fr/privacy_faq.html#toc-terms-personal-info. Accessed 28 Aug 2010
43. Google—Centre de confidentialité, Principes applicables à la confidentialité (2010), Available: http://www.google.com/intl/fr/corporate/privacy_principles.html. Accessed 28 Aug 2010
44. Y. Duan, J. Canny, Protecting user data in ubiquitous computing: towards trustworthy environments. LNCS (Springer, Berlin, 2005)
45. V. Bellotti, What you don't know can hurt you: privacy in collaborative computing, in *British Computer Society Conference on Human-Computer Interaction* (1996), pp. 241–261
46. Organization for Economic Co-operation and Development (OECD), Guidelines on the protection of privacy and transborder flows of personal data (1980), Available: http://www.oecd.org/document/18/0,3343,en_2649_34255_1815186_1_1_1_1,00.html. Accessed 6 May 2010
47. Organization for Economic Co-operation and Development (OECD), At crossroads: personhood and digital identity in the information society. The working paper series of the OECD directorate for science, technology and industry (2008), Available: http://www.oecd.org/LongAbstract/0,3425,en_2649_34223_40204774_119684_1_1_1,00.html. Accessed 21 May 2010
48. European Union, Directive 95/46/EC of the European Parliament and of the Council (1995), Available: http://eur-lex.europa.eu/LexUriServ/LexUriServ.do?uri=CELEX:31995L0046:en:HTML. Accessed 26 March 2010
49. Wikipedia, Data protection directive (2011), Available: http://en.wikipedia.org/wiki/Data_Protection_Directive. Accessed 26 March 2010
50. P. Brown, Privacy in an age of terabytes and terror. Sci. Am. Mag. **299**, 46–47 (2008)
51. Overview of the Privacy Act of 1974, United States Department of Justice's Office of Privacy and Civil Liberties (OPCL), 2010
52. R. Agrawal et al., Hippocratic databases, in *Proceedings of the 28th International Conference on Very Large Data Bases (VLDB)*, Hong Kong-China 2002, pp. 143–154
53. J.J. Luna, How to be invisible: a step-by-step guide to protecting your assets, your identity, and your life. New York, 2012
54. CSA Model Code for the Protection of Personal Information, The Canadian Standards Association, 2000
55. Personal Information Protection and Electronic Documents Act, Government of Canada—Department of Justice, 2000
56. The Canadian Standards Association, Privacy code, Available: http://www.csa.ca/cm/ca/en/privacy-code. Accessed 12 May 2010
57. Privacy Act, 1985, Available: http://laws-lois.justice.gc.ca/eng/acts/P-21/
58. The Japanese Act on the Protection of Personal Information, The Government of Japan, 2003
59. Hewlett-Packard, Japan Personal Information Protection Act (JPIPA)—legislative summary, Available: http://h71028.www7.hp.com/ERC/cache/571127-0-0-0-121.html. Accessed 12 May 2010
60. ZL Technologies Inc., Regulation overview of the Japanese Act on the protection of personal information, Available: http://www.zlti.com/resources/docs/Rules%20and%20Regulations/ZL.RR.Japan-PIPA.pdf. Accessed 12 May 2010
61. Australian Government—Office of the Privacy Commissioner. Information Privacy Principles under the Privacy Act 1988—Plain English Summary, Available: http://www.privacy.gov.au/materials/types/law/view/6892. Accessed 12 May 2010
62. L. Baeriswyl, G. Mangeat, Protection des données personnelles sur Internet: la situation en Suisse (2000), Available: http://www.juriscom.net/chr/2/ch20000621.htm. Accessed 10 Nov 2011
63. Commission nationale de l'informatique et des libertés (CNIL), Security guide of personal data (2010), Available: http://www.cnil.fr/fileadmin/documents/en/Guide_Security_of_Personal_Data-2010.pdf. Accessed 10 Nov 2011

64. La loi de la protection des données à caractère personnelles de la Tunisie, 2004
65. United States Department of Health & Human Services, Office for civil right privacy brief: summary of the HIPAA privacy rule (2003), Available: http://www.hhs.gov/ocr/privacy/hipaa/understanding/summary/privacysummary.pdf. Accessed 28 May 2010
66. Wikipedia, The Health Insurance Portability and Accountability Act (HIPAA) of 1996 (2011), Available: http://en.wikipedia.org/wiki/Health_Insurance_Portability_and_Accountability_Act#Privacy_Rule. Accessed 26 April 2010
67. Wikipedia, The Gramm–Leach–Bliley Act (GLB) (2011), Available: http://en.wikipedia.org/wiki/Gramm%E2%80%93Leach%E2%80%93Bliley_Act. Accessed 26 April 2010
68. G. Ben Ayed, S. Ghernaouti-Hélie, Privacy requirements specification for digital identity management systems implementation: towards a digital society of privacy, in *6th International Conference for Internet Technology and Secured Transactions (ICITST-2011)*, Abu Dhabi, UAE, 2011
69. M. Rotenberg, The privacy law sourcebook 2004: United States law, international law, and recent developments: Electronic Privacy Information Center, 2004

Part II
Interoperability Through Service-Orientation

Chapter 5
DigIdeRP Framework

The proper study of mankind is the science of design.
Herbert Simon (American scientist, 1916–2001)

We cover in this chapter three main areas. The first area deals with the foundations and basic concepts of service orientation and service-oriented architecture. The second one deals with a high level and detailed descriptions of DigIdeRP framework. In the last area, we present and describe each block that composes SoaML-based DigIdeRP framework.

5.1 Privacy Implementations: Current Landscape

The experts remarked that, in general, regulators are often a step behind fast-paced digital innovations, so concerns over privacy and data protection are important examples [1]. Though much has been done since the 1970s for developing legal principles and provisions for the protection of privacy, this has led to the growth of a number of so-called privacy-enhancing technologies (PETs) with the aim of giving users greater control over their personal data. These can be thought of as falling into three categories: protecting privacy through proxy; using P3P, protecting privacy through the absence of traceability (e.g. the Freenet Project[1]). In addition to the privacy-enhancing systems, improvements in cryptography have been contributing to the growing security of data (e.g. PKI). The use of PETs has been limited in the digital world, thus, the market for privacy is still relatively small. Besides, most consumers find that the available systems are too complex or burdensome to apply properly. Others lack awareness relating to the possibility of privacy violations [2]. However, the common thing here is that privacy is dealt

[1] http://freenetproject.org

G. Ben Ayed, *Architecting User-Centric Privacy-as-a-Set-of-Services*, Springer Theses,
DOI: 10.1007/978-3-319-08231-8_5, © Springer International Publishing Switzerland 2014

from a technical perspective and lacks multidisciplinary and integrated approach. Particularly, privacy is approached broadly and has not been specified for digital identity. Projects such as Kentara Initiative[2] are basically digital identity projects and may cover few project-specific elements of privacy. Service orientation and privacy are implemented mainly from managerial perspectives such as [3] and how enterprises insure a certain level of privacy combined within Service Level Agreement (SLA).

5.2 Service-Oriented Architecture

Service-Oriented Architecture (SOA) improves the solution construction and benefits the enterprise as a whole. The author [4] points eight areas: (1) the cost and effort of cross-application integration is significantly lowered when applications being integrated are SOA-compliant; (2) service-orientation promotes the design of services that are inherently reusable; (3) composing existing services into aggregate services could reduce processing overhead and skill-set requirements; (4) leveraging the legacy investment through the participation in service-oriented integration architectures. The cost and effort of integrating legacy and contemporary solutions is lowered; (5) the cost and effort of application development is reduced after a proliferation of standardized XML data representation; (6) since SOA can centralize inter-application and intra-application communication as part of standard IT infrastructure, the cost of scaling communications infrastructure is reduced; (7) SOA establishes a vendor-neutral communications framework, it frees IT departments from being dependent to a single proprietary development and/or middleware platform; and (8) the cost and effort to respond and adapt to business or technology-related change is reduced.

The term ‘architecture’ is employed in different expressions to refer to different meanings. We don’t intend here to define the term but to list few common usages in the field. ‘Application architecture’ is to an application development team what a blueprint is to a team of construction workers. An ‘enterprise architecture’ specification is to an organization what an urban plan is to a city. When coupled with ‘architecture’, service-orientation takes on a technical connotation, thus SOA can refer to application architecture or the approach used to standardize technical architecture across the enterprise [4]. The same author defines SOA with his own words as “a form of technology architecture that adheres to the principles of service-orientation. When realized through the Web services technology platform, SOA establishes the potential to support and promote these principles throughout the business process and automation domains of an enterprise” [4]. However, the same author adds that there is no official set of service-orientation principles but eight common set of principles most associated with

[2] https://kantarainitiative.org/

Table 5.1 Comparison of object-oriented and service-oriented design approaches

	Service-orientation	Object-orientation
Processing logic	Creation of activity-agnostic services that are driven into action by messages	Creation of objects bound with data
Dependencies	Loose coupling between services	Predefined class dependencies resulting in more tightly bound objects
Interfaces	Coarse-grained interfaces (service descriptions) and messages contain as much information as possible for the completion of a given task	Fine-grained interfaces (APIs). Tasks are performed through RPC/API calls
Scope	Significant variation in scope	Small and specific in scope
States	The creation of services to remain as *stateless* as possible	The creation of more *stateful* objects
Composition	Composition and orchestration of services	Composition and inheritance among objects

service-orientation: (1) services are reusable; (2) services share a formal contract that describes each service and defines the terms of information exchange; (3) services are loosely coupled; (4) services abstract underlying logic and the only thing that is visible to the outside world is what is exposed via the service contract; (5) services are composable and may compose other services; (6) services are autonomous and has full control within its boundary and is not dependent on other services in term of execution and governance; (7) services are stateless and should not be required to manage state information; and (8) services are discoverable services should allow their descriptions to be discovered and understood by humans and service requestors.

In Table 5.1, we present a comparison that is provided by Erl [4] between aspects of object-oriented and service-oriented design approaches. We should notice that when designing a software application, even if we opt for service oriented approach, we still need to use object oriented concepts. In fact, in response to the question: is your team was building a service-oriented architecture (SOA)? Erl [4] answers "my architect thinks it's service-oriented, my developers insist it's object-oriented, and my analysts wish it would be more business-oriented. All I can tell you is that it isn't what it was before we started building Web services". In his answer, the project manager provides different perceptions to SOA given by different project members to stress that SOA should be seen in a broader view. A technical architecture that comprises of Web services is a common but dangerous assumption that leads to the number one mistake made by organizations intending to adopt SOA [4].

The author [4] illustrates the need of service orientation in order to enrich distributed systems and environments. Relating to reality, he explains that in an average cosmopolitan city, people have service-oriented businesses. Individual companies are service-oriented in that each provides a distinct service that can be used by multiple consumers. Collectively, these businesses comprise a business community. It makes sense for a business community not to be served by a single business outlet providing all services. By decomposing the community into specialized, individual outlets, we achieve an environment in which these outlets can be distributed.

5.3 High-Level View Description of DigIdeRP Framework

Security is pervasive through the entire cycle of information processing and the design of security systems requires the adoption of design best practices in order to reduce risks [5]. Particularly, we recognize that technology alone cannot guarantee resolution for the concerns surrounding a multi-facets and complex issue of digital identity-related privacy. An interdisciplinary and integrated approach should be adopted in order to reduce identity-related privacy breaches and harms. It is demonstrated that technology and technical solutions are not enough to tackle privacy issues. To ensure the right solution, we should take into account laws, societal norms, markets, privacy policies, fair information practices, procedures and technology to guide the implementation of the system [6–8]. Building digital identity management systems should be in accordance of DigIdeRP requirements on which implementers could build the system from requirements engineering phase. Consequently, privacy should be engineered and integrated from the start, rather than attaching it after the fact [9].

The purpose of DigIdeRP framework helps to align digital identity-related privacy projects initiatives with the organization's business goals and security strategy. The author [10] highlights that DigIdM systems initiatives must be approached from a strategic point of view with a high level of clarity on objectives. The framework should carefully consider and clearly define business goals, strategy, policies and standards, along with detailed identity management architecture, specifications and a road map. DigIdeRP framework focuses primarily on disassembling DigIdeRP requirements into services that can integrate a SOA. DigIdM systems, or identity systems as illustrated in Fig. 5.1, are hosted either inside an enterprise or across enterprises. Within distributed systems, cloud computing and SOA, networked identity systems could collaborate through calls of a set of orchestrating services. More specifically, identity services would collaborate with digital identity-related privacy services. Through the use of DigIdeRP Framework, Enterprise/Information System security team that brings together IT security architects, designers, developers, and analysts, may be able to disassemble DigIdeRP requirements into autonomous, granular and loosely coupled set of services and build Privacy-as-a-Set-of-Services (PaaSS). The available services enable on-demand privacy; whenever a party is in need of one or multiple elements of DigIdeRP, he could invoke the associated service or services to respond to his need. The privacy service-orientation would inevitably resolve complexities and issues associated with different and various DigIdeRP implementations within identity systems.

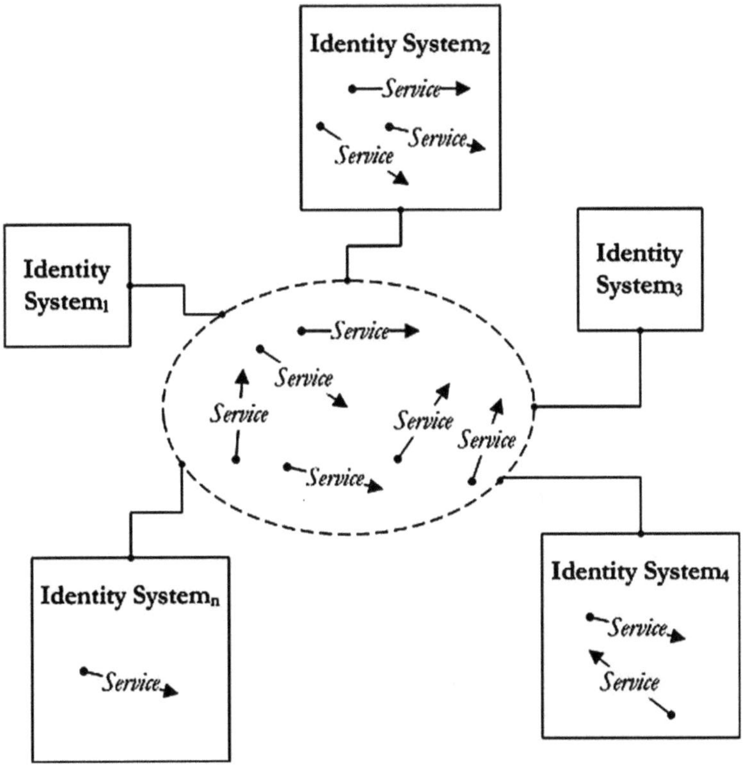

Fig. 5.1 DigIdeRP as a set of service

More specifically, the target system encloses a set of services: in-rest and in-action represent the possible states of the services. The first one is represented by a horizontal arrow and it refers to an inactive and a ready-to-use service that is still not invoked by a party. The second one is represented by a vertical arrow and it refers to an active service that is invoked by a participant. In this state, a negotiation and communication channel is established between the parties: service sender and service receiver(s). If the negotiation is a success, the service is consumed and if it is a failure, the service is released. The dash line eclipse delimits services hosting environment that could be any machine, set of distributed machines, or cloud computing environment (see Fig. 5.2).

The eclipse represents also the circle-of-trust among the participants: subject, service provider (SP), and identity provider (IdP) within digital identity federation. The services descriptor directory system plays the role service discovery system, which describes all available services in term of a service's objective, a detailed description, a hosting system address, constraints, etc. and allows service access by the participants. Services choreographies describe the cooperation between available services, more specifically between service interfaces, to respond to participants' needs.

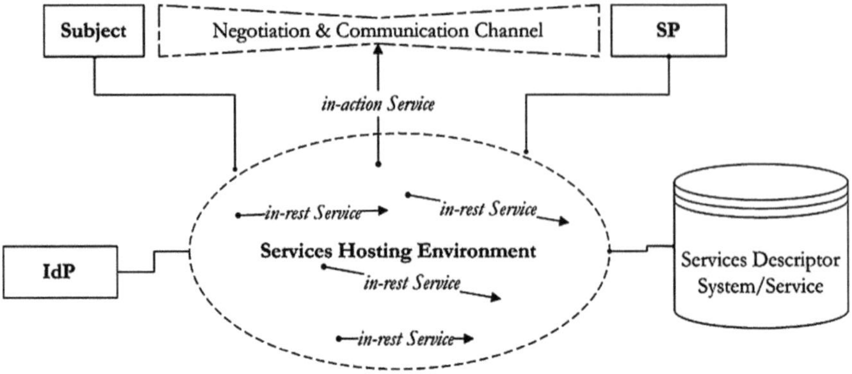

Fig. 5.2 DigIdeRP services

We suggest a five-layer framework that would help IS security tem to implement DigIdeRP requirements into a set of services that can accommodate any SOA. DigIdeRP Framework will serve as a basis for vital understanding between business management and technical managers on digital identity related privacy initiatives. The framework relays on the idea that privacy requirements should be taken into consideration from the beginning of the identity system development project. For this reason, 'privacy by design', 'privacy from the outset or 'privacy from the start' are introduced [9]. However, if we consider privacy from the beginning of the project, how could we turn DigIdeRP requirements into design, architectures and then implementations? DigIdeRP Framework is an answer to this question. We identified from works of [10–13], DigIdeRP Framework that govern five layers in order to implement the target system PaaSS (Fig. 5.3). We borrow the argument of [12] when he describes his layered framework: "these layers are roughly analogous to a network protocol stack with a many-to-many relationship between successive layers and most certainly do not imply a top-down waterfall-style software engineering process". The framework presents layers as an ordered sequence, however, in practice, there is an iterative process to assure that each layer supports effectively and enforces requirements of the adjacent ones.

The framework provides five practical steps as a basis of identity management project roadmap. Each layer is composed by a set of specific activities. Specifically, the framework is divided into five layers and three mapping gateways: (1) purpose-level SOA is concerned with articulating the purpose and motivations of the project. The purpose, context and motivation of the implementation initiative should be established, clearly stated, and supported by the organization executives and management level. System's purpose specification should be the first step in designing any identity system [6]. The implementation/engineering vision should be defined and how it could accommodate to the business strategic vision. We encourage specifying the vision and purpose with partners and understanding together business requirements such as policies, regulations, trust, and

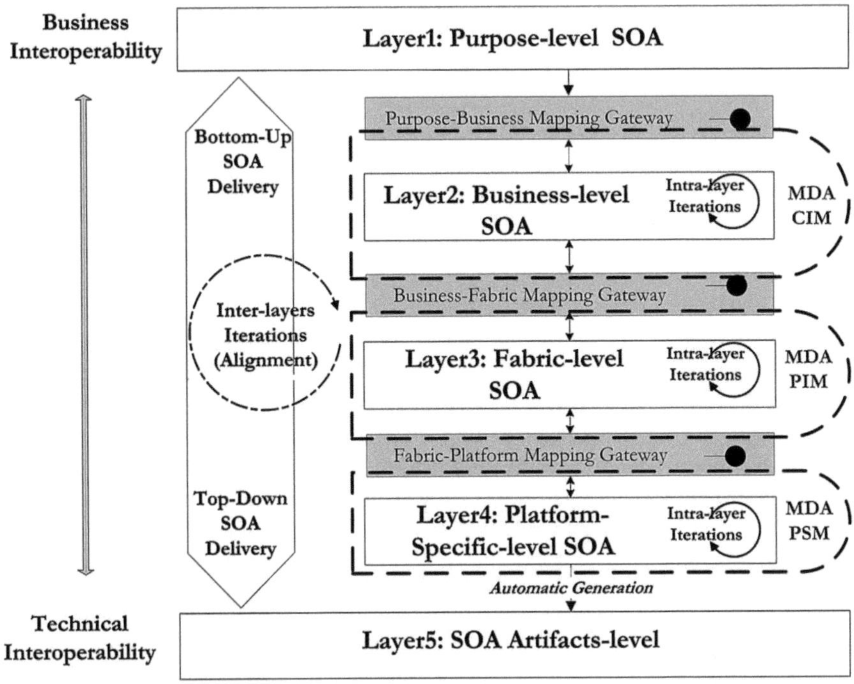

Fig. 5.3 Layered DigIdeRP framework

dependencies. The purpose is to build digital identity-related Privacy-as-a-Set-of-Services that could accommodate any SOA. More specifically, the focus of the framework is not to develop SOA, rather developing privacy services that could be hosted within SOA. It is designing services for SOA and not designing an SOA itself. In the purpose-business mapping gateway, we look for sources such as privacy policies, procedures, fair information practices and project-specific needs in order to identify DigIdeRP requirements further in the next level; (2) business-level SOA deals with specifying clear DigIdeRP requirements and taking into consideration DigIdM architectural and technical models constraints. In the business-fabric mapping gateway, we identify service candidates' pool from DigIdeRP requirements. Thus, the mapping gateway will facilitate and ease the transition between the two layers; (3) fabric-level SOA copes with identifying and specifying the services, conversation and collaboration between them (interfaces and choreographies), and the way of calling them. In the fabric-platform mapping gateway, we consider several services' deployment environment constraints in the service design such as the component diagrams in UML2 through which we model the transition from business software architecture into technical software architecture; (4) platform-specific-level SOA handles with specific-platform deployment environment of the services such deployment diagram in UML2 that depicts a static view of the run-time configuration; and (5) SOA artifacts-level in which we

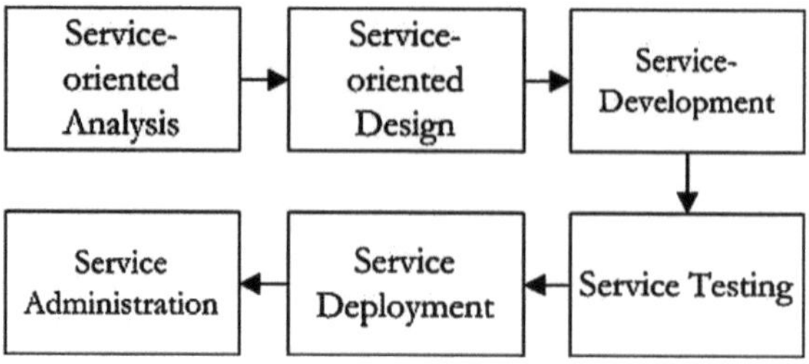

Fig. 5.4 Common phases of an SOA delivery lifecycle

generate through ready-to-use automatic transformation rules, implementations and codes of SOA artifacts. Completeness of services' implementation that is generated on this level depends on the maturity of the layer 4 outputs. We could evolve DigIdeRP Framework to be fully in accordance of model-driven engineering (MDE)/model-driven architecture (MDA) approach. However, we consider purpose-business mapping gateway, business-level SOA, and business-fabric mapping gateway as key elements of MDA Computational-Independent Model (CIM); business-fabric mapping gateway, fabric-level SOA, and fabric-platform mapping gateway as key elements of MDA Platform-Independent Model (PIM); and fabric-platform mapping gateway and platform-specific-level SOA as key elements of MDA Platform-Specific Model (PSM).

Inter- and intra-DigIdeRP layers iterations are consequence of SOA delivery lifecycle and strategies alignment. Erl [4] presents in Fig. 5.4 the common phases of an SOA delivery lifecycle. In the service-oriented analysis stage, we determine the potential scope of the SOA and identify the service candidates. The service-oriented design is a heavily standards-driven phase that incorporates industry conventions and service-orientation principles into the service design process. In this stage, we define business processes as services orchestration. The service development phase is the construction one and then services are required to undergo rigorous testing prior to deployment into a production environment. Finally, after deploying services, services monitoring, messages tracking and management, performance management come to the forefront. The same author points that lifecycle stages should be organized into a process that can support a transition toward an SOA in order to fulfill project requirements. More specifically, a strategy is needed to make the transition within a given budget and timeline. The strategy must be based on an organization's priorities to establish the correct balance between the fulfillment of long-term migration goals and short-term requirements. Three common strategies have emerged: (1) top-down strategy process steps includes define relevant enterprise-wide ontology, align relevant business models (including entity models) with new or revised ontology, perform service-oriented analysis, perform service-oriented design, develop the required services, test the services and all service operations, and deploy the services; (2) bottom-up strategy process steps includes service-oriented analysis, service-oriented design, service development, test the services, and deploy the services; and (3) agile or meet-in-the-middle strategy process steps includes top-down analysis, focusing first on key parts of the ontology and related business entities, however, when the top-down analysis has sufficiently progressed, perform service-oriented analysis, service-oriented design, develop, test, and deploy the services. As far top-down analysis continues to progress, revisit business services [4]. Based on the need of adopting the right delivery strategy and previous descriptions, we provide in Table 5.2 a summary of a comparison between of SOA delivery approaches. A set of criteria are identified.

In DigIdeRP Framework, we choose the combination of top-down and bottom-up strategies in a different way from agile approach. The agile strategy allows for the business-level analysis to occur concurrently with service design and development. In the framework, business-level analysis starts to occur in accordance of top-down strategy without going through service design and implementation. Thus, intra-layer iterations occur. However, as far as the top-down progresses, we adopt on-demand bottom-up strategy to allow going back from layers to upper ones. Thus, inter-layers iterations occur.

5.4 OMG Service-Oriented Modeling Language

SoaML is a 'in Process' specification from the Object Management Group (OMG), which describe a UML profile and metamodel for designing services within a service-oriented architecture. The specification extends UML2 to support the activities of service modeling and design and to fit into an overall model-driven development approach. Among others, SoaML goals are: (1) identifying services, functional capabilities, requirements and dependencies between the services; (2) defining service consumers and providers; (3) and policies for using and providing services. Particularly, the profile and metamodel accommodate different perspectives: service consumer perspective, service provider perspective, and system design perspective and describe consumers requirements, providers offerings and the interaction and agreements between them. In addition, SoaML specifications provide definitions of a service and SOA. The service is defined as "an offer of value to another through a well-defined interface and available to a community" and SOA as "an architectural paradigm for defining how people, organizations and systems provide and use services to achieve results" [14, 15].

SoaML is chosen for multiple reasons: (1) SoaML is a modeling language that helps to ensure an easy understanding and validation by the project members since SoaML permits a technology-neutral representation of the services; (2) SoaML supports the activities for modeling service that could be accommodated by service oriented architecture. SoaML permits to identify service candidates and to design services for SOA and not SOA itself; (3) SoaML fits into an overall model-driven development approach, which is considered as an important aspect because MDA facilitates the design when requirements change; (4) SoaML enables business oriented and systems oriented services architectures to mutually and collaboratively support organization's mission [14]; and (5) SoaML contains

Table 5.2 SOA delivery approaches comparison

	Top-down	Bottom-up	Agile
Approach nature	'Analyst-first' approach	Services delivery on an 'as needed' basis	A combination of the two approaches
Approach motivation	Incorporating service-oriented design principles into business analysis	Services as a means of fulfilling integration requirements by building web services	The strategy allows for the business-level analysis to occur concurrently with service design and development
Pre-requisites	Business requirements have already been collected and defined	Business requirements have already been collected and defined	–
Number of steps	7	5	7
Advantages	High quality service architecture and adaptable services delivery	Create easily Web services as required by applications	Employing this strategy fulfills both short and long-term needs
Drawbacks	Significant investment (time and money) in analysis without showing immediate results	Service-orientation principles are rarely considered	The agile strategy is more complex than the two others because it needs to fulfill two opposing sets of requirements. Maintenance is hard and costly

modeling constructs that would help to identify service candidates from DigIdeRP requirements and processes. Reducing and decomposing requirements into a set of service candidates, is still an open issue.

5.5 Detailed View of SoaML-Based DigIdeRP Framework

DigIdeRP initiatives are to be approached from a strategic point of view with a high level of clarity on objectives. The framework considers and clearly defines business goals, strategy, policies and standards, along with detailed system's architecture, specifications and a road map. In other words, framework helps to align DigIdeRP initiatives with the organization's business goals and security strategy. The blocks in the framework determine a roadmap that security team could follow to successfully implement PaaSS. The framework will serve as a basis for vital understanding between business management and technical managers on all DigIdeRP initiatives.

We detail the blocks in each layer of the framework. In the purpose-level SOA layer (Fig. 5.5), we articulate the need of implementing digital identity-related Privacy-as-a-Set-of-Services. In the purpose-business mapping gateway, we identify the privacy requirements sources related to digital identity such as policies, fair information practices, laws and procedures. We classified them into three groups: (1) privacy business-specific requirements represent the privacy requirements related to identity in particular industry or filed such as banking, health, and education; (2) privacy domestic requirements represent the recurring privacy needs and recommendations related to identity presented by national bodies and local privacy authorities; and (3) privacy global requirements represent the recurring privacy needs and polices related to identity presented by international bodies, regional policy-makers and global legal framework. In addition, it encompasses also the requirements that are neither domestic nor business-specific and the practices and assessment tools that are provided by organizations having a global vision. More details can be found in Chap. 4. In business-level SOA layer, we specify four blocks: (1) functional requirements' specification. DigIdeRP requirements are already specified in Chap. 4; (2) DigIdM technical model. The technical models are already been covered and compared in Chap. 3. DigIdM identity federation is elected because it secures distributed systems and allows to better preserve privacy; (3) DigIdM deployment perspective. ITU report [16] classifies DigIdM systems' works and projects into a landscape of three perspectives: (a) network operator centric perspective in which capabilities that maximize and protect network assets are sought; (b) application service provider centric perspectives in which capabilities that maximize and protect application assets are sought; and (c) user-centric perspective in which capabilities that allow privacy protection and user control over digital identity are sought. Considered as a derivate of DigIdM identity federation, user-centric identity federation is a novel and promising approach that provides more control over digital identity [17]. That's why

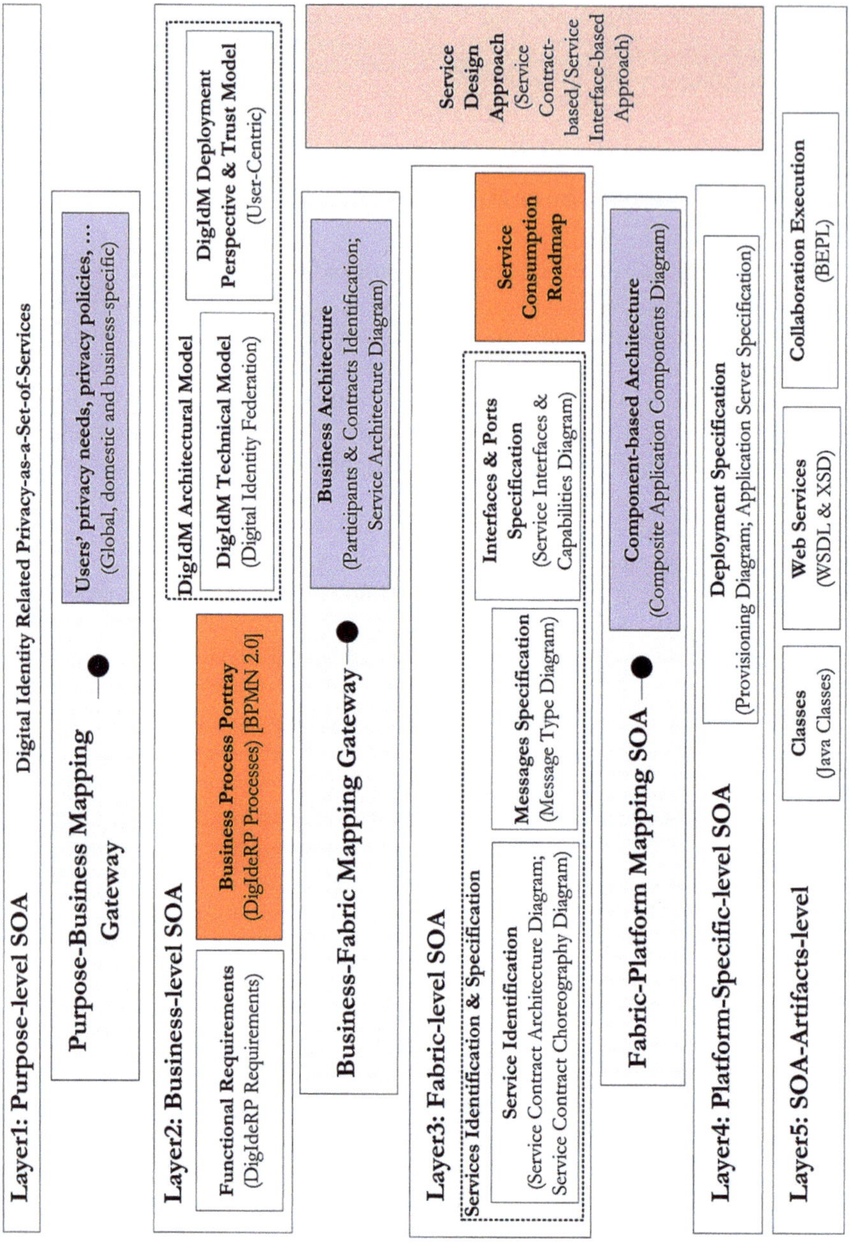

Fig. 5.5 Detailed DigIdeRP framework

user-centric approach will be adopted in the framework. DigIdM technical model and DigIdM deployment perspective blocks are grouped into DigIdM architectural model envelope; and (4) business process portray.

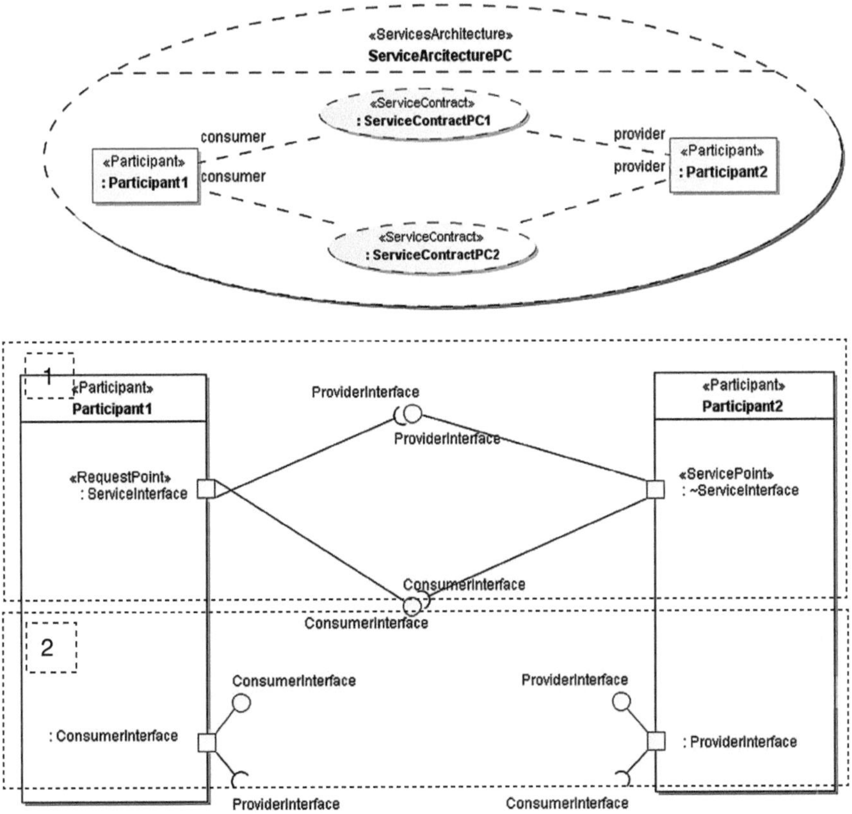

Fig. 5.6 Service design approaches

5.6 Service Design Approaches

Service design approach is an inter-layers block. SoaML modeling capabilities support the service "contract-based" and "interface-based" approaches, which follow the "ServiceContract" and "ServiceInterface" elements [18]. We have to choose between the two approaches before undertaking activities in the business-fabric mapping gateway, fabric-platform mapping gateway, layer 3, and layer 4. In Fig. 5.6, we illustrate the main differences between the service-contract and service-interface approaches. In the upper part of the Fig. 5.5, service architecture diagram, we present two service contracts between two participants: participant 1 and participant 2.

In the second part of Fig. 5.6, the envelope 1 represents the design of interfaces in alignment of the service-interface approach. The service interface plays the role of an intermediary between the consumer and provider interfaces. In the respect of the interface specification, we will have an interface conjugate (represented by ~)

which will play the role of the second party of the conversation. A service channel is set between two interfaces to play the role of communication channel and it is represented as both consumer interfaces and provider interfaces are close to each others. Through service point, services are delivered and through request point, services are consumed. However, in the service-contract approach, envelope 2 of Fig. 5.6, the interfaces communicate directly without intermediaries and the service channel is represented logically through intefaces' names: ConsumerInterface and ProviderInterface. Moreover, service-contract approach requires an already established business and collaboration agreement between parties. In the adopted DigIdM identity federation technical model, CoT sets the agreement between parties of the identity federation, thus, service-contract approach is the best-fit in our context. The diagram part, envelope 2, will not be included since consumer and provider interfaces' names are shown in composite application component diagram.

5.7 Business Process-Based Portray: DigIdeRP Processes

Organizations may initially have invested in separate Business Process Management (BPM) and Service-Oriented Architecture (SOA) initiatives but these concepts go hand-in-hand and ultimately the convergence between them will help organizations achieve greater value. BPM enables organizations to perform and manage end-to-end business processes. Meanwhile, SOA facilitates the decoupling of reusable business logic embedded in IT system assets into business services and it enables access to business services through industry standard interfaces. BPM orchestrates the end-to-end business processes and the invocation of services, which in turn may call other services to automate steps in the process. The synergy between BPM and SOA not only allows users better control of the business process as it decoupled from the IT architecture, but also better alignment between Business and IT [19].

PaaSS system will be fully designed in accordance to DigIdeRP requirements. We chose to describe DigIdeRP requirements in flow chart-based notation: Business Process Modeling Notation (BPMN 2.0) for three major reasons: (1) process-based description enriches the requirements; (2) after identifying services, process-based description will provide the way how services will be consumed and invoked in order to fully execute the process (see Fig. 5.7); and (3) to exploit, in future work, the compelling synergies between BPM and SOA, and to explore intersections between BPMN and SoaML. We detail the six processes that we built up: (1) **ServiceRequest** process: the subject sends a service request to the SP pool ServiceRequestInfo, which encloses service name crafted with subject's identifier. The SP sends back to the subject SP-Subject identity request. Subject sets up a negotiation context in order to reach an agreement over the terms of digital identity contract. The subject demands to the SP to draft the contract detailing access, use, collection, dissemination, disclosure, destruction, and modification purposes,

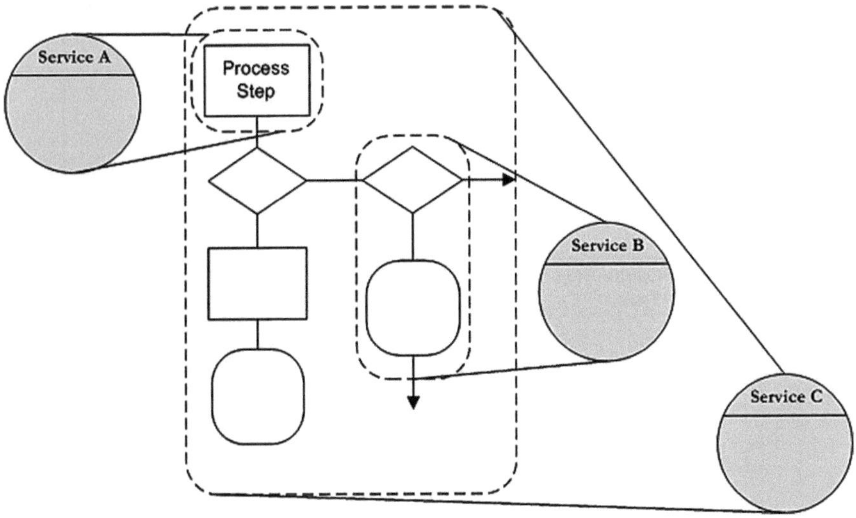

Fig. 5.7 Business processes decomposition and services consumption

permissions and rights over digital identity attributes. The SP provides the contract to the subject. The latter may accept all the terms of the contract or decline (few of) them and express his intention to renegotiate the contract's terms until both of them reach the desired agreement. The subject sends the IdP(s) specification that the SP should in touch with in order to receive the desired attributes. Sending IdP specification implies giving a proof of agreement over all terms of digital identity contract. Therefore, SP sends digital identity request to the specified IdP(s), which in turn send(s) digital identity contract to the subject and ask(s) him his consent before sending digital identity attributes to the SP. The subject checks and compare on-hand contract with the one sent by each IdP and makes the appropriate decision. The subject could either reject and therefore sends decision revocation over send service request or send his acceptance and therefore each IdP sends digital identity attributes to the SP. The SP receives from each IdPs digital identity attributes, releases the service and sends service availability note to the subject, who finally consumes it; (2) **ProfileToChallenge** process: the subject sends a profile-to-challenge-request to the SP in order to be able to access his profile, check its validity and have the capability to change it. The SP sends the possessed profile that is drawn from digital identity attributes aggregation (Fig. 5.8). The subject may send a change, update or modify profile request to the SP, which confirms the update operation. However, no action will be undertaken if the subject is in agreement with his profile; (3) **Enrollment** process: the subject sends an enrollment request with digital identity attributes to the IdP, which saves attributes and confirms a successful subject's enrollment; (4) **PeriodicDigitalIdentityToUpdate** process: IdP sends periodic digital identity to-update-request to the subject in order to check whether digital identity attributes are still valid or should be updated. The subject receives his record and may change

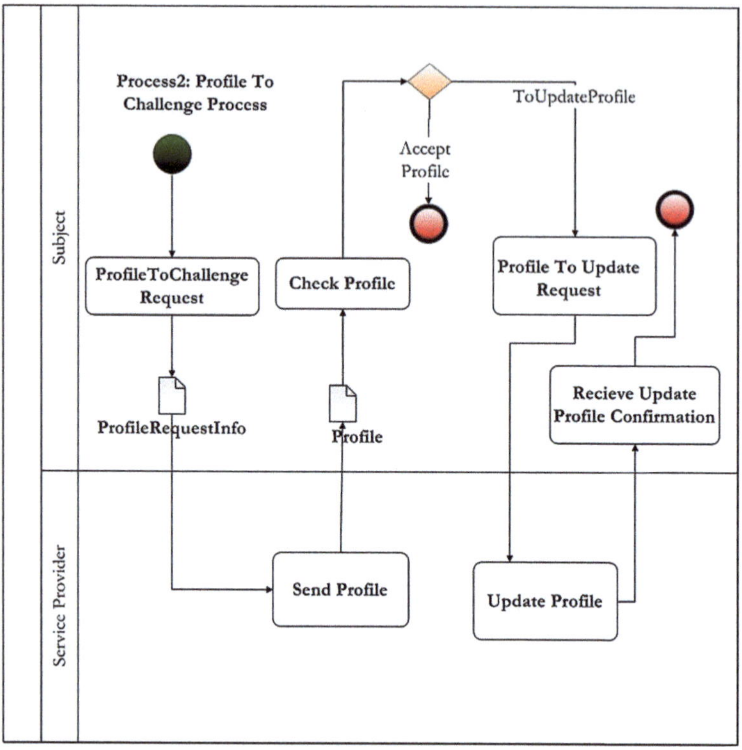

Fig. 5.8 ProfileToChallenge process BPMN description

it; (5) **DigitalIdentityToUpdate** process: The subject sends a request to the IdP in the sake of updating the values of digital identity attributes that are in is hold. The IdP updates the record and sends an update confirmation back to the subject; and (6) **EditDigitalIdentity** process: The subject sends an EditDigitalIdentityRequestInfo to IdP, which sends back to Subject digital identity attributes.

5.8 Business Architecture

In the Fig. 5.9, we represent the participants that we identified (subject, IdP, SP) with "participant" stereotype. The participant participates in a service contract with a specific role, which may change when participating in other service contracts. The dash lines and labels represent the roles (consumer, provider) of each participant in the service architecture. We provide an overview of each service contract: (1) ContractAgreement service contract: the subject plays the role of a Sender of ContractAgreement to the SP in order to establish an agreement about contract's conditions and provisions. The SP plays the role of a Receiver of ContractAgreement;

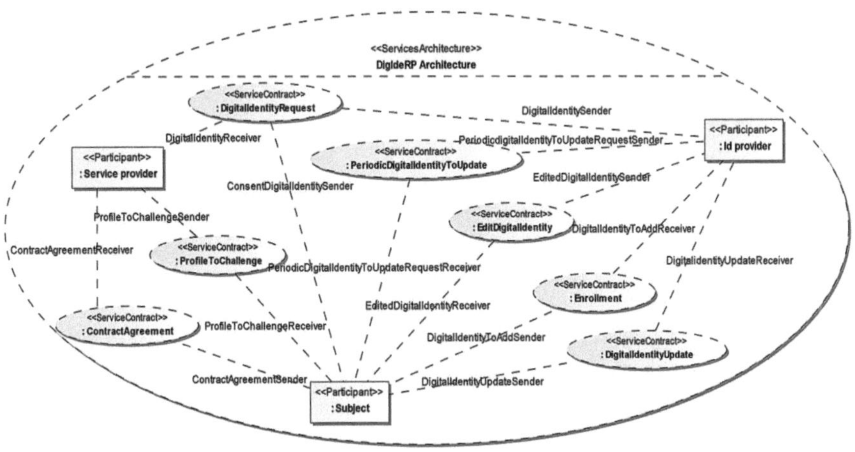

Fig. 5.9 DigIdeRP services

(2) ProfileToChallenge service contract: the Subject has the ability to challenge his profile that it is in hold by the SP. Thus, the Subject plays the role of a Sender and the SP as a Receiver; (3) DigitalIdentityRequest service contract involves all participants: the Senders are the Subject and IdP; and the Receiver is SP. The SP asks to receive specifications of IdP(s), which transfer to the SP the subject's identity attributes. The transfer is not possible if the subject has not clearly given his consent. The Subject conveys to the IdP the consent about digital identity attributes dissemination to the SP; (4) PeriodicDigitalIdentityToUpdate service contract involves the IdP, which is the in-charge of the timing process, as a Sender of the attributes' update request to the Receiver: Subject; (5) DigitalIdentityUpdate service contract allows to describe the Subject's ability to send new digital identity attributes' values to the IdP, which plays the role of the Receiver; (6) Enrollment service contract involves the Subject as a Sender of digital identity to be added request to the IdP, which receives the request; and (7) EditDigitalIdentity service contract implicates the IdP as a sender of digital identity attributes to the Subject, who is already expressed willing to edit digital identity.

5.9 Service Identification and Specification

In this section, we show and explain how to disassemble DigIdeRP requirements, enriched with BPMN process-based description into a set of seven services. Service architecture diagram, Fig. 5.9, shows seven contracts, which specify services without regard for their implementations. For each service, we provide details through establishment of SoaML service contract architecture diagram, service contract choreography diagram, and message type diagram. Each service contract diagram shows though a connector that an interaction is established between two roles stereotyped "consumer" and "provider".

Methods are available either in consumer service interface or provider service interface. The latter can invoke methods that are available through consumer service interface and vice versa. The service choreography diagram highlights the negotiation and communication process between service interfaces in term of calls of methods. Moreover, different inputs of the methods are messages that are described in messages diagrams. We've identified a set of seven services: (1) ContractAgreement service: (2) DigitalIdentityRequest service; (3) DigitalIdentityToUpdate service; (4) PeriodicDigitalIdentityToUpdate service; (5) Enrollment service; (6) ProfileToChallenge service; and (7) EditDigitalIdentity service.

We provide a detailed description of three out of seven services as an illustration how we can use the framework. The **ContractAgreement** service could be called by the subject in order to set an agreement with SP on the purpose of identity collection, handling (retention) duration, disclosure, and access capabilities. In the ContractAgreement service contract diagram, "usage" relations are specified between the roles ContractAgreementSender and ContractAgreementReceiver. "Roles" are becoming "interfaces" that hold available methods visible to each others. We illustrate through choreography diagram the usage relations and how each interface may invoke methods. The provider invokes toServeRequest method in order to have the authorization to use a service such as an online payment service. The method requires as an input toServe message that encloses the serviceName (Fig. 5.10). The consumer invokes SP-SubjectDigitalIdentityRequest method to request the digital identity from the provider. The provider invokes the ContractAgreementRequest method requesting to establish a mutual agreement on terms of the contract about how digital identity will be maintained. The consumer invokes sendContract method with Contract message type that encloses the purpose of attributes' usage and identityRetentionDuration of attributes. When the contract is received, the provider decides whether he agrees or declines. If he agrees, the provider invokes the ContractAgreement IdPSpec method with termsOfContract, IdPRef, agreementConfirmation, and SubjectRef parameters in order to send IdP specification to the consumer; else he demands to change terms of the contract until they reach mutual agreement by invoking rectificationContractRequest method with termsOfContract and agreementConfirmation as input parameters.

The second service is **DigitalIdentityRequest**. The service contract involves two kinds of roles: two providers and a consumer are bounded through connectors. Each role is represented by an interface that comprises all available methods. Methods are visible and could be used by any of the three interfaces, (Fig. 5.11).

The consumer calls SP-IdPDigitalIdentityRequest method as a requesting expression of desired digital identity attributes. For this reason, a copy of the contract bound with SPRef is to be sent through the method to the provider: DigitalIdentitySender. The latter in his turn calls consentRequest method requesting the consent of the other provider: ConsentDigitalIdentitySender. The latter, may accept and freely send his consent by invoking identityDissemination method with consentToSend and SPRef inputs. Then, the DigitalIdentitySender provider sends digital identity attributes to the consumer by calling the sendDigitalIdentity method.

The consumer replies and delivers the service to the ConsentDigitalIdentitySender by invoking serviceDelivery method with ToServe message type. The

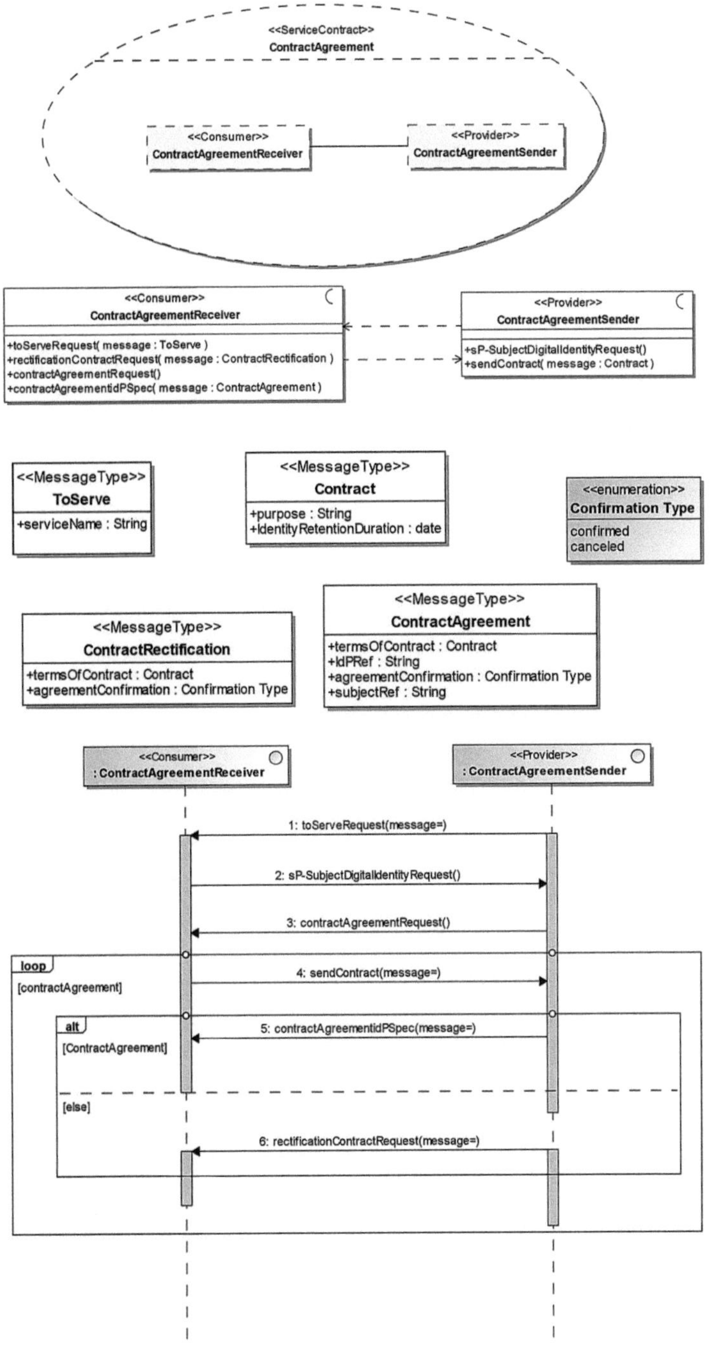

Fig. 5.10 ContractAgreement service contract, message type and choreography diagrams

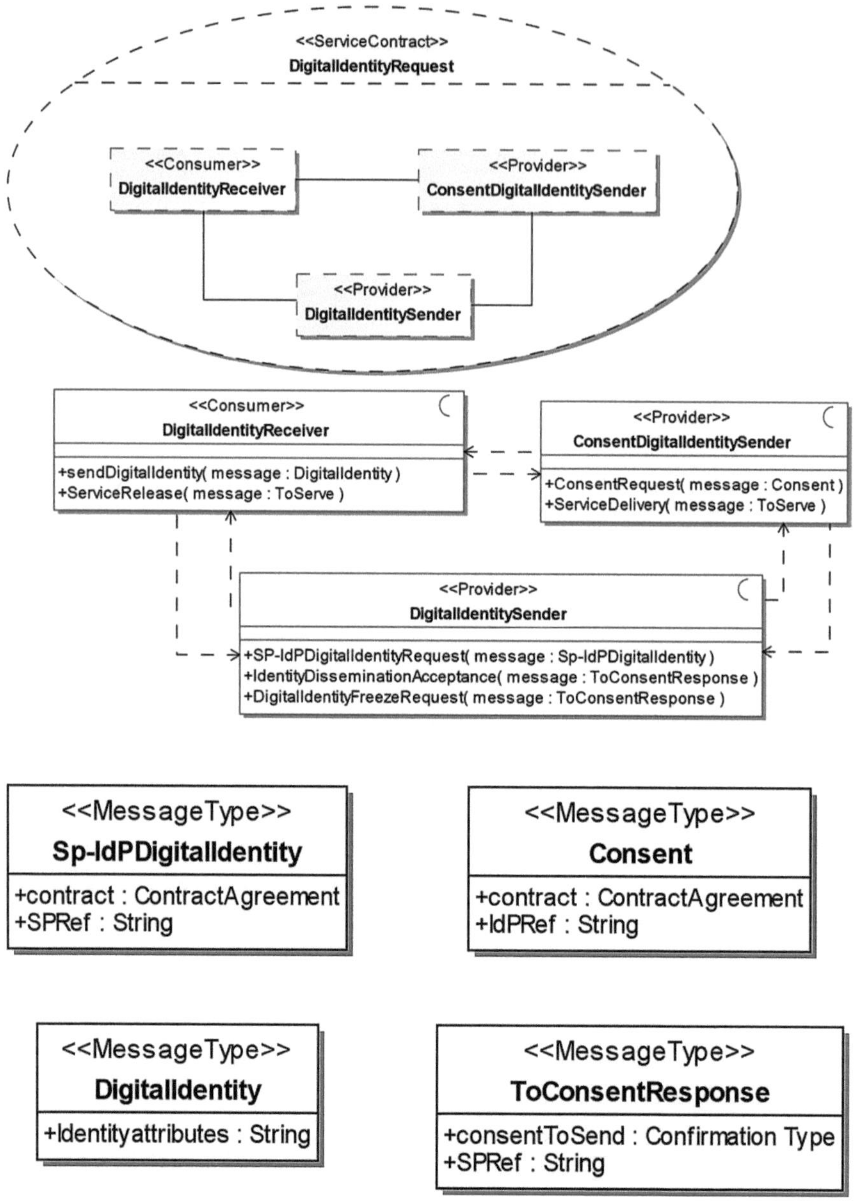

Fig. 5.11 DigitalIdentityRequest service contract, message type and choreography diagrams

ConsentDigitalIdentitySender provider may also decline to send the consent by calling digitalIdentityFreezeRequest method with the parameter ToConsentResponse that comprises the response to the consent request: consentToSend, and the ID of SP: SPRef. Therefore, ConsentDigitalIdentitySender provider releases the service request

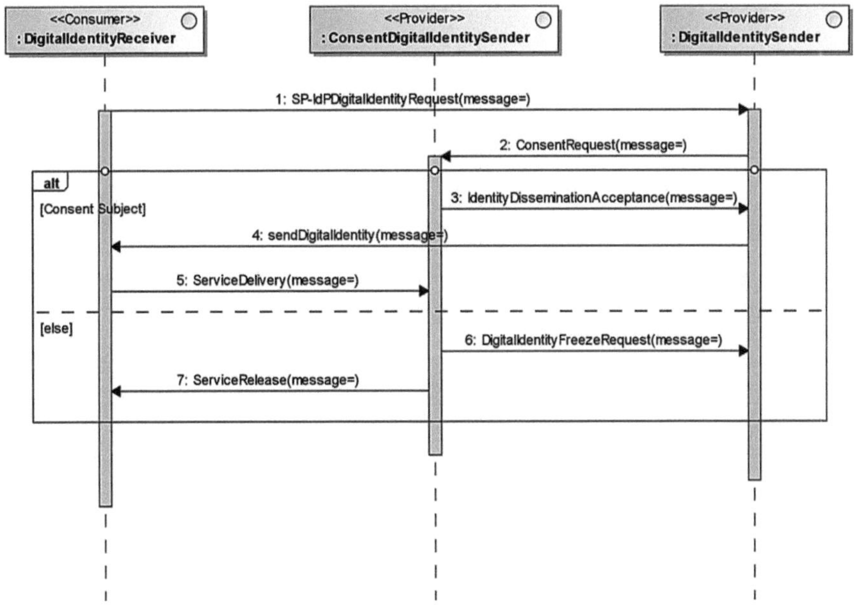

Fig. 5.11 continued

and asks to not be served through calling the serviceRelease method, (Fig. 5.11). In the **ProfileToChange** service, (Fig. 5.12), the service contract is set between the consumer ProfileToChallengeReceiver and the provider ProfileToChallengeSender. Each role is represented by an interface. The consumer invokes ProfileRequest with profileProperties message type, which encloses subjectRef information. The provider invokes sendProfile method with profile message type. The consumer is able to send a request for a profile change by invoking profileToUpdateRequest method with profile properties message type. The provider receives a profile change acknowledgement as a result of consumer's invocation of updateProfileConfirmation method with UpdatedProfileConfirmation message type.

5.10 Service Consumption Roadmap

We provide few an implementation of how we could execute the processes by invoking the identified services. In other terms, we translate processes into sevice(s) call(s). For ProfileToChallengeprocess, we call these services: (1) (Service Name: ProfileToChallenge Service, Requester: Subject, Recipient: SP, Signal: ProfileRequest); (2) (Service Name: ProfileToChallenge Service, Requester: SP, Recipient: Subject, Signal: SendProfile); (3) (Service Name: ProfileToChallenge Service, Requester: Subject, Recipient: SP, Signal: ProfileToUpdateRequest); and (4) (Service Name: ProfileToChallenge Service, Requester: SP, Recipient:

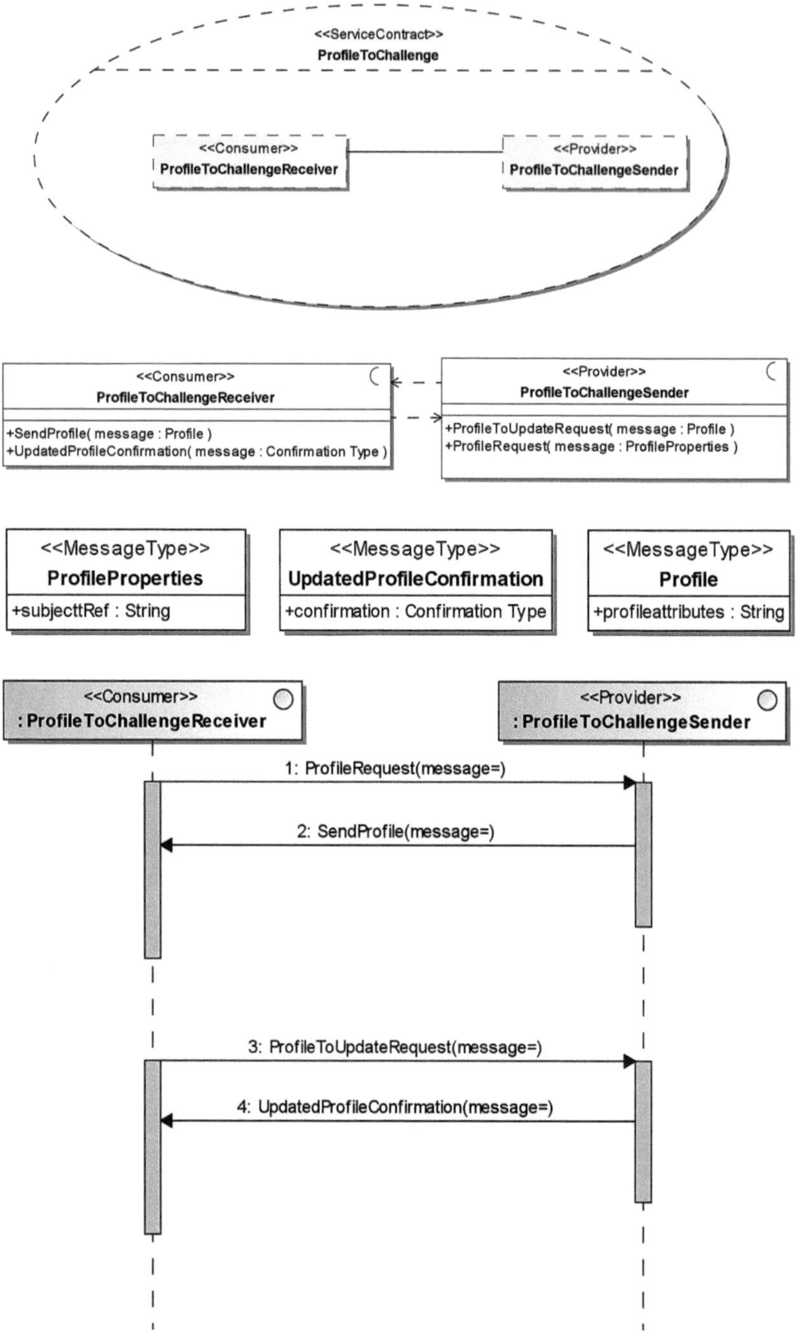

Fig. 5.12 ProfileToChallenge service contract, message type and choreography diagrams

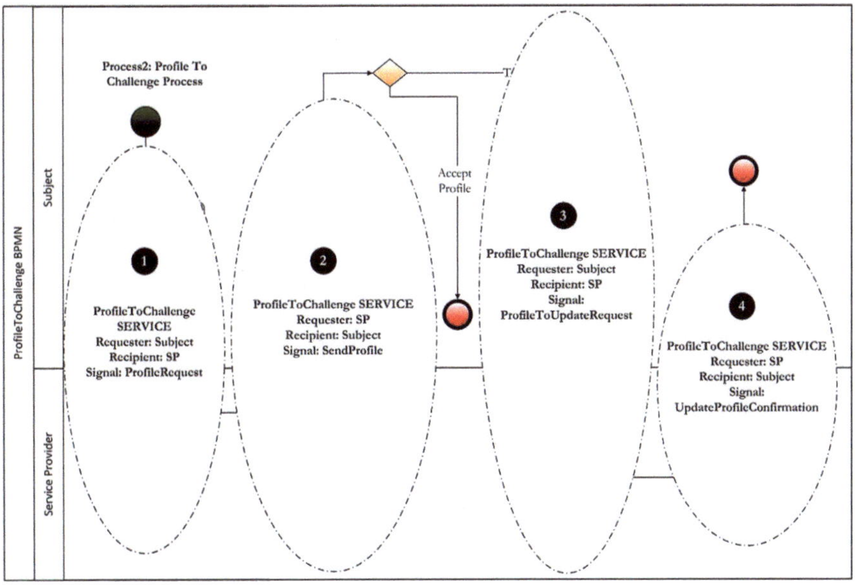

Fig. 5.13 Services consumption diagram of ProfileToChallenge process

Subject, Signal: UpdateProfileConfirmation). In Fig. 5.13, we provide the service(s) calls map. For **DigitalIdentityToUpdate** process, we call 1) (Service Name: PeriodicDigital Identity ToUpdate Service, Requester: IdP, Recipient: Subject, Signal: PeriodicDigitalIdentityToUpdateRequest); (2) (Service Name: PeriodicDigital Identity ToUpdate Service, Requester: Subject, Recipient: IdP, Signal: SendUpdatedIdentity); and (3) (Service Name: PeriodicDigital Identity ToUpdate Service, Requester: IdP, Recipient: Subject, Signal: PeriodicDigitalIdentityToUpdate Confirmation).

5.11 Component-Based Architecture

In the pre-implementation step, we provide and describe, through composite application component diagram (Fig. 5.14), different components to be implemented further. The square-shaped elements represent ports that each of them encloses consumer and ProviderInterfaces. Service points and request points are represented by stereotypes and small squares in different components. Service points are drawn from service contract diagram provider's role and the request point from consumer's role. For instance, in ContractAgreement service contract diagram, the consumer ContractAgreementReceiver is translated into ContractAgreementReceiver request point and the provider ContractAgreementSender is in its turn translated into service point ContractAgreementSender. Rather than we represent explicitly

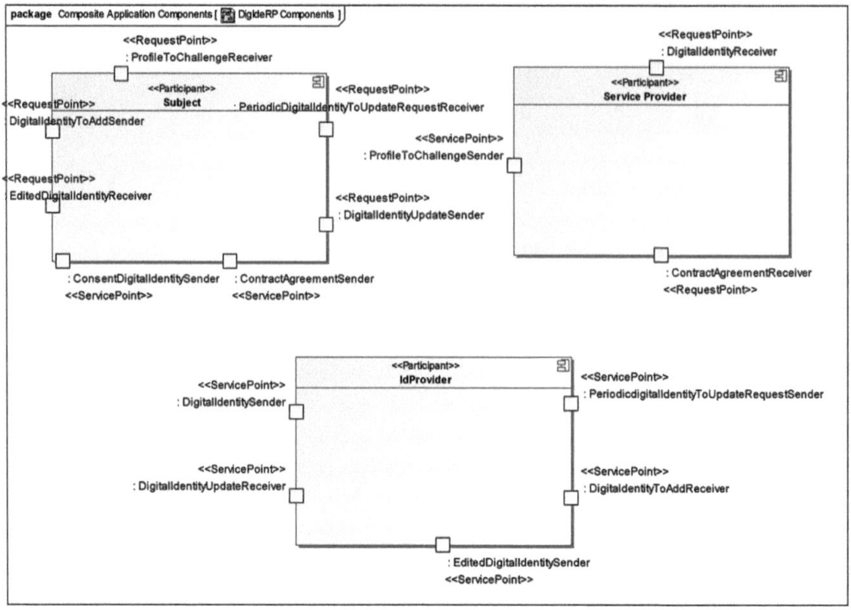

Fig. 5.14 DigIdeRP composite application component diagram

channels in composite application component diagram, we specify "logical" service channels through interfaces' names because we adopted SoaML service contract-based approach. For example, the subject component may invoke available methods in SP component through ContractAgreementReceiver request point; and vice versa SP component could invoke available methods in subject component through service point ContractAgreementSender. Each component has a set of both service and/or request points: (1) Subject has {ContractAgreementSender, and ConsentDigitalIdentitySender} service points and {DigitalIdentityUpdateSender, EditedDigitalIdentityReceiver, DigitalIdentityToAddSender, PeriodicDigitalIdentityToUpdateRequestReceiver, and ProfileToChallengeReceiver} request points; (2) SP has {ProfileToChallengeSender} service point and {DigitalIdentityReceiver, and ContractAgreementReceiver} request points; and (3) IdP has {DigitalIdentitySender, DigitalIdentityUpdateReceiver, PeriodicDigitalIdentityToUpdateRequestSender, DigitalIdentityToAddReceiver, and EditedDigitalIdentitySender} service points.

5.12 Deployment Specification

The composite application component diagram is a platform-independent diagram; however, the provision diagram (Fig. 5.15) is a platform-dependent one. Here is Java Enterprise Edition, JEE-dependant provision diagram. The three components

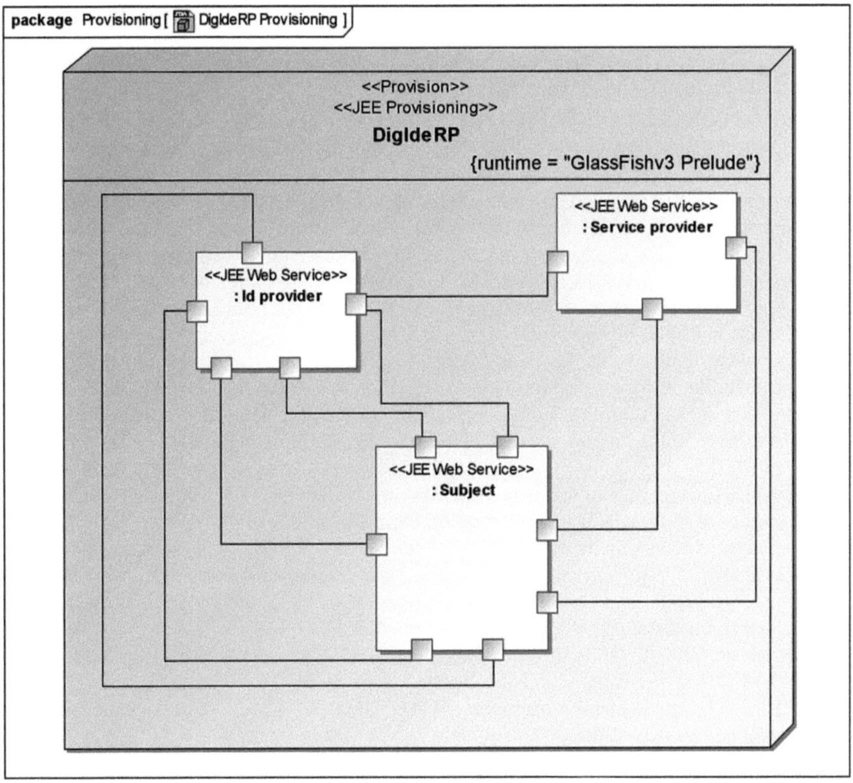

Fig. 5.15 DigIdeRP provisioning diagram

that are enclosed in DigIdeRP composite application component diagram are transformed into three JEE Web services that are available within Glass Fish JEE application server. We can envisage other application server such as JBoss or we can distribute the Web services into different types of application servers. The square-shaped elements represent again the ports. The links between Web services represent the possibility and available opportunity of methods invocations between interfaces through ports.

References

1. H. Noonan (ed.), *Identity*. Stanford Encyclopedia of Philosophy (2009)
2. International Telecommunication Union (2006), *Digital life*. ITU Internet Report Available: http://www.itu.int/osg/spu/publications/digitalife/docs/digital-life-web.pdf. Accessed 21 May 2010
3. G. Karjoth et al, Privacy-enabled services for enterprises, in *13th International Workshop on Database and Expert Systems Applications (DEXA'02)*, 2002, p. 483

4. T. Erl, *Service-Oriented Architecture: Concepts, Technology, and Design.* (Prentice Hall, New Jersey, 2005)
5. M. Benantar, *Access Control Systems: Security, Identity Management and Trust Models.* (Springer, Berlin, 2006)
6. Center for Democracy and Technology (2007), *Privacy principles for identity in the digital age* (Draft for comment—version 1.4). Available: http://www.cdt.org/files/pdfs/20071201_IDPrivacyPrinciples.pdf. Accessed 28 May 2010
7. Organization for Economic Co-operation and Development (OECD) (2008), At crossroads: personhood and Digital Identity in the Information Society. The working paper series of the OECD directorate for science, technology and industry. Available: http://www.oecd.org/LongAbstract/0,3425,en_2649_34223_40204774_119684_1_1_1,00.html. Accessed 21 May 2010
8. L. Lessig, *Code and Other Laws of Cyberspace.* (Basic Books, New York, 2000)
9. A. Cavoukian (2009), *Privacy by Design.* Available: http://www.privacybydesign.ca/pbdbook/PrivacybyDesignBook.pdf. Accessed 7 June 2010
10. S. Vanamali, Identity management framework. Inf. Syst. Control J. **4**, 49–52 (2004)
11. C. Daeseon et al, An information security model for the next generation application service, in *Proceedings of the 2nd International Workshop for Asia Public Key Infrastructure*, Taipei, Taiwan, 2002
12. R. Sandhu, Engineering authority and trust in cyberspace: the OM-AM and RABC way, in *Proceedings of the 5th ACM Workshop on RBAC*, 2000, pp. 111–119
13. J. Sherwood, Opening up the enterprise. Comput. Secur. J. **19**, 710–719 (2000)
14. OMG (2009), *Service oriented architecture modeling language (SoaML)—specification for the UML profile and metamodel for services (UPMS).* Available: http://www.omg.org/spec/SoaML/1.0/Beta2/PDF/. Accessed 20 June 2010
15. D. Krishnan (2009), *OMG Releases Draft of SoaML.* Available: http://www.infoq.com/news/2009/01/omg-releases-soaml. Accessed 22 June 2010
16. ITU Focus Group on Identity Management (FG IdM), *Report on identity management use cases and gap analysis*, 2007
17. A. Jøsang, S. Pope, User-centric identity management, in *Proceedings of the AusCERT Asia Pacific Information Technology Security Conference*, 2005, pp. 1–6
18. T. Varvarigou, V. Andronikou, Identity management in GRID computing and service oriented architectures: research and practice. Paper presented at the Second Multidisciplinary Workshop on Identity in the Information Society (IDIS 2009), London, UK, 2010
19. Price Waterhouse Coopers Thought Leadership Institute (2010), *From the white board to the bottom line: the case for pursuing process maturity through business process management.* Available: http://download.pwc.com/ie/pubs/from_the_white_board_to_the_bottom_line.pdf. Accessed 18 Mar 2011

Chapter 6
SOA-Artifacts-Level: Implementation of Privacy-as-a-Set-of-Services

A quite simple, but powerful, technology is that empowers individuals to keep control over and manage their digital identities.

Dave Winer (American software developer and web writer)

6.1 SoaML Design Toolkit

We use MagicDraw UML (version 16.5) software with Cameo SOA+ extension (version 16.5) to design the system with SoaML diagrams. We integrated Eclipse IDE (version 3.4) with ModelPro SDK (version 1.1) in order to generate the code of SOA-related artifacts, including Java code for service interfaces and SCA components, and XSD, WSDL, SCA Composite, and BPEL specification files (Fig. 6.1).

Supporting SoaML, ModelDriven.org ModelPro is an open source MDA provisioning engine that is able to produce a wide variety of artifacts from models. ModelPro is able to produce executable web service implementations for services architectures defined in SoaML [1].

6.2 SOA Artifacts Related to the Service Provider Participant

Figure 6.2 shows the list of Service Provider SOA artifacts that are generated in the form of Web services contract artifacts (WSDL and XSD files) and Java files (JEE project). The red and discontinued lines shows in which Java files API JAXWS annotations are included. All the numbers that are labeling XSD and Java files correspond to the numbers that are included in the header of the codes listed in Sect. 6.5.

G. Ben Ayed, *Architecting User-Centric Privacy-as-a-Set-of-Services*, Springer Theses,
DOI: 10.1007/978-3-319-08231-8_6,

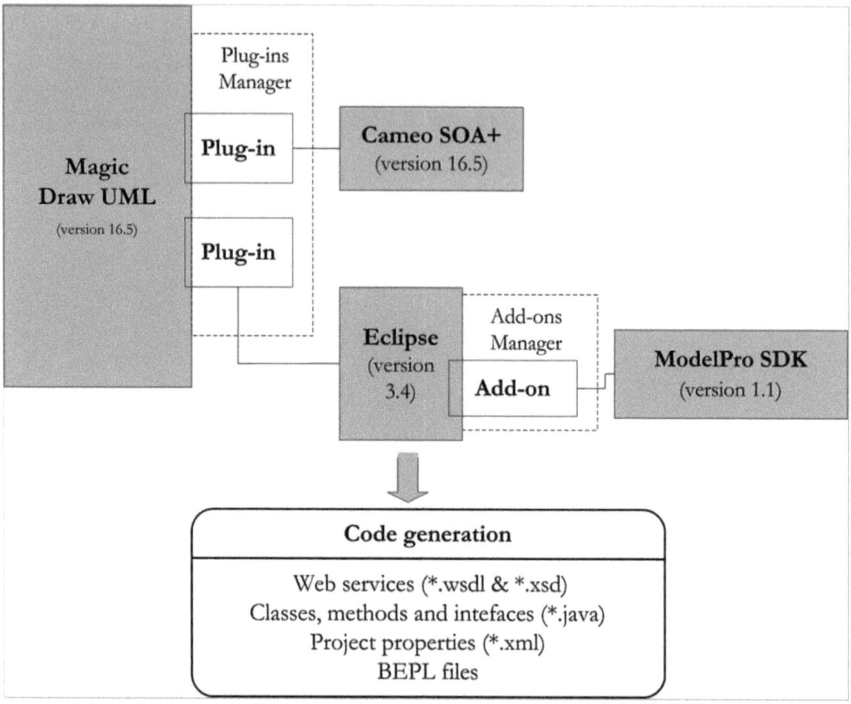

Fig. 6.1 SOA tools

6.3 SOA Artifacts Related to the Identity Provider Participant

Figure 6.3 shows the list of Identity Provider SOA artifacts that are generated in the form of Web services contract artifacts (WSDL and XSD files) and Java files (JEE project). The red and discontinued lines shows in which Java files API JAXWS annotations are included. All the numbers that are labeling XSD and Java files correspond to the numbers that are included in the header of the codes listed in Sect. 6.5.

6.4 SOA Artifacts Related to the Subject Participant

Figure 6.4 shows the list of Subject SOA artifacts that are generated in the form of Web services contract artifacts (WSDL and XSD files) and Java files (JEE project). The red and discontinued lines shows in which Java files API JAXWS annotations are included. All the numbers that are labeling XSD and Java files correspond to the numbers that are included in the header of the codes listed in Sect. 6.5.

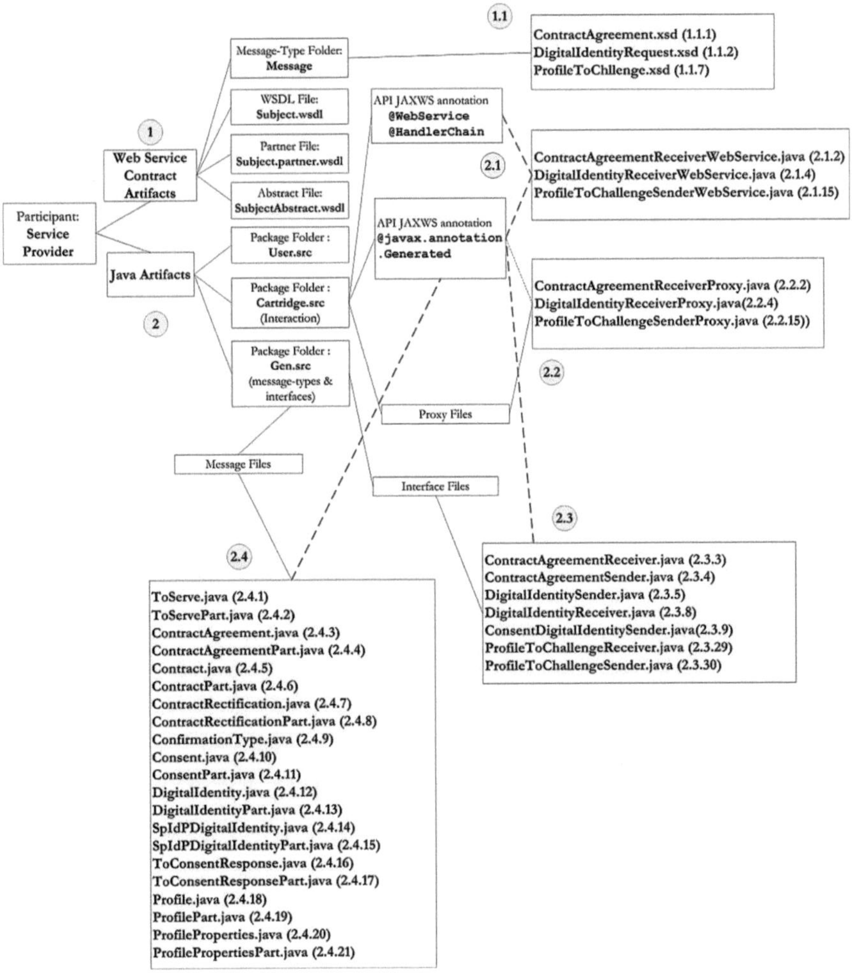

Fig. 6.2 SP SOA artifacts

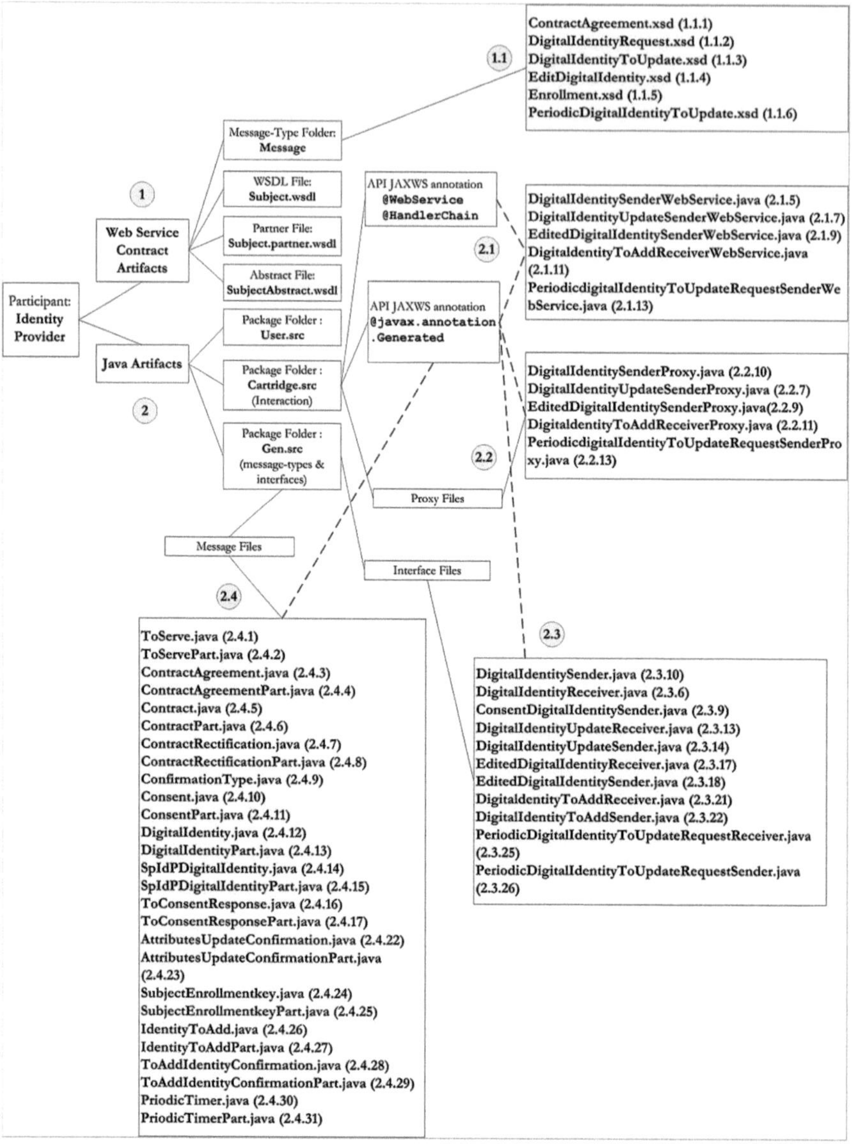

Fig. 6.3 IdP SOA artifacts

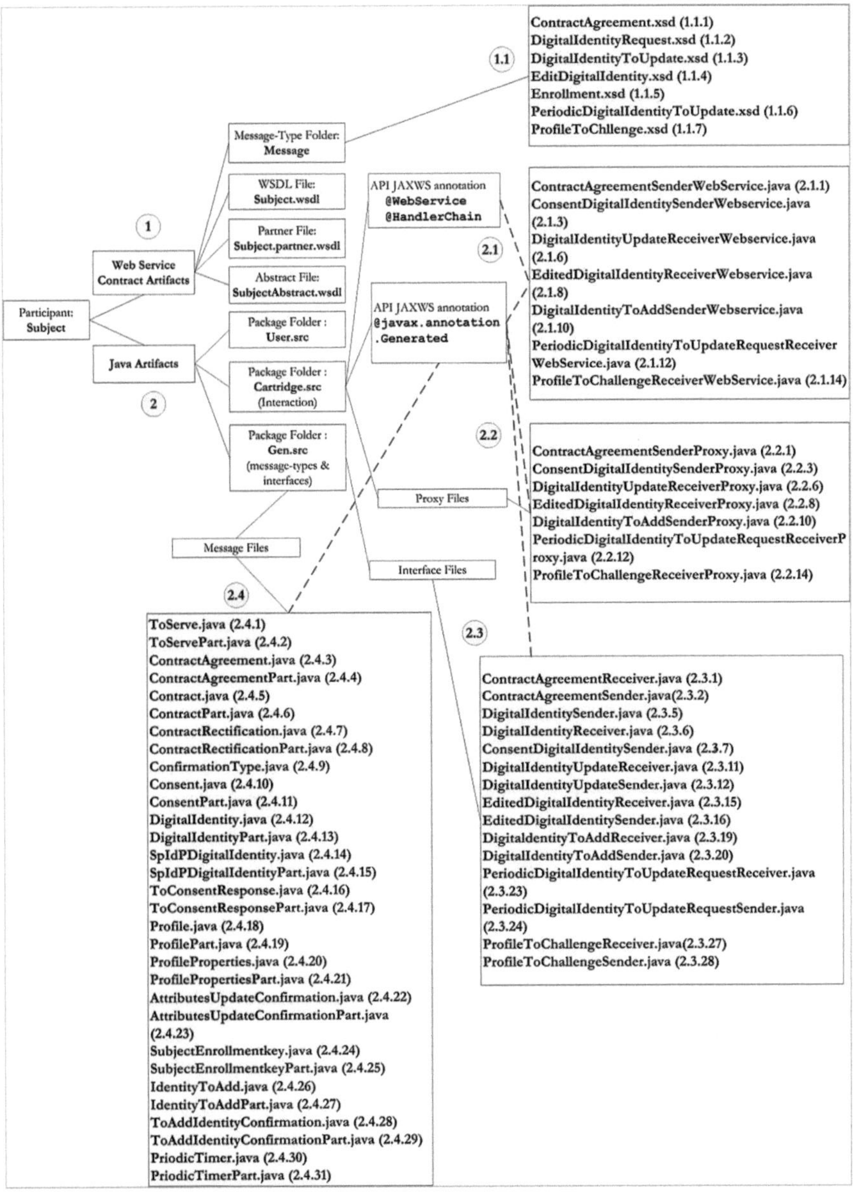

Fig. 6.4 Subject SOA artifacts

6.5 SOA Artifacts Code Generation

A short list of codes skeletons in Java and XSD are generated and numbered in the head of the following codes. The numbers allow establishing links between codes that are generated versus SOA artifacts codes that have to be generated, see Figs. 6.2, 6.3 and 6.4.

```
++++++++++++++++++++++++++++++++++++++++++++++++++++++++++++++++
++++++++++++++++++++++++++++++++++++++++++++++++++++++++++++++++
++++++++++++++++++++++++++++++++++

                  Participant: Subject, SP, IdP
          Message type: ContractAgreement.java (2.4.3)

++++++++++++++++++++++++++++++++++++++++++++++++++++++++++++++++
++++++++++++++++++++++++++++++++++++++++++++++++++++++++++++++++
++++++++++++++++++++++++++++++++++

@javax.annotation.Generated(value = { "ModelPro SoaML
Cartridge" }, date = "2011-10-25T14:52:36.366+01:00")
public interface ContractAgreement {
      public
org.modeldriven.examples.message.contractagreement.Contract
getTermsOfContract();
            public void setTermsOfContract(
org.modeldriven.examples.message.contractagreement.Contract
_value);
      public String getIdPRef();
      public void setIdPRef(String _value);
      public
org.modeldriven.examples.message.contractagreement.Confirmati
onType getAgreementConfirmation();
      public void setAgreementConfirmation(
org.modeldriven.examples.message.contractagreement.Confirmati
onType _value);
      public String getSubjectRef();
      public void setSubjectRef(String _value);}
```

```
++++++++++++++++++++++++++++++++++++++++++++++++++++++++++++
++++++++++++++++++++++++++++++++++++++++++++++++++++++++++++
++++++++++++++++++++++++++++++++
```

Participant: Subject, SP, IdP
Message type: ContractAgreementPart.java (2.4.4)

```
++++++++++++++++++++++++++++++++++++++++++++++++++++++++++++
++++++++++++++++++++++++++++++++++++++++++++++++++++++++++++
++++++++++++++++++++++++++++++++

@javax.annotation.Generated(value = { "ModelPro SoaML
Cartridge" }, date = "2011-10-25T14:52:36.366+01:00")
public class ContractAgreementPart implements Serializable {
        private
    org.modeldriven.examples.message.contractagreement.Contract
    termsOfContract;
        public
    org.modeldriven.examples.message.contractagreement.Contract
    getTermsOfContract() {
            return termsOfContract;}
        /*Set the value of termsOfContract*/
        public void setTermsOfContract(
        org.modeldriven.examples.message.contractagreement.Cont
    ract _value) {
            this.termsOfContract = _value;}
        /*Get the value of IdPRef*/
        private String idPRef;
        public String getIdPRef() {   return idPRef;}
        /*Set the value of IdPRef*/
        public void setIdPRef(String _value) {
            this.idPRef = _value;}
        /*Get the value of agreementConfirmation*/
        private
    org.modeldriven.examples.message.contractagreement.Confirmati
    onType agreementConfirmation;
        public
    org.modeldriven.examples.message.contractagreement.Confirmati
    onType getAgreementConfirmation() { return
    agreementConfirmation;}
        /*Set the value of agreementConfirmation*/
        public void setAgreementConfirmation(
        org.modeldriven.examples.message.contractagreement.Conf
    irmationType      _value) {this.agreementConfirmation =
    _value;}
        /*Get the value of subjectRef */
        private String subjectRef;
        public String getSubjectRef() {return subjectRef;}
        /*Set the value of subjectRef*/
        public void setSubjectRef(String _value)
        {this.subjectRef = _value;}}
```

```
+++++++++++++++++++++++++++++++++++++++++++++++++++++++++++
+++++++++++++++++++++++++++++++++++++++++++++++++++++++++++
+++++++++++++++++++++++++++++++
```

Participant: Subject, SP, IdP
Message type: Consent.java(2.4.10)

```
+++++++++++++++++++++++++++++++++++++++++++++++++++++++++++
+++++++++++++++++++++++++++++++++++++++++++++++++++++++++++
+++++++++++++++++++++++++++++++

@javax.annotation.Generated(value   =   {   "ModelPro   SoaML
Cartridge" }, date = "2011-10-25T14:52:36.366+01:00")
public interface Consent {
      public
org.modeldriven.examples.services.contractagreement.ContractA
greement getContract();
      public void setContract(
org.modeldriven.examples.services.contractagreement.ContractA
greement _value);
public String getIdPRef();
public void setIdPRef(String _value);}


+++++++++++++++++++++++++++++++++++++++++++++++++++++++++++
+++++++++++++++++++++++++++++++++++++++++++++++++++++++++++
+++++++++++++++++++++++++++++++
```

Participant: Subject, SP, IdP
Message type: ConsentPart.java(2.4.11)

```
+++++++++++++++++++++++++++++++++++++++++++++++++++++++++++
+++++++++++++++++++++++++++++++++++++++++++++++++++++++++++
+++++++++++++++++++++++++++++++

@javax.annotation.Generated(value = { "ModelPro SoaML
Cartridge" }, date = "2011-10-25T14:52:36.366+01:00")
public class ConsentPart implements Serializable {
      private
org.modeldriven.examples.services.contractagreement.ContractA
greement contract;
      public
org.modeldriven.examples.services.contractagreement.ContractA
greement getContract() {          return contract;}
      public void setContract(
org.modeldriven.examples.services.contractagreement.ContractA
greement _value) {          this.contract = _value; }
      private String idPRef;
      public String getIdPRef() {   return idPRef;    }
      public void setIdPRef(String _value) {this.idPRef =
_value;     }}
```

```
++++++++++++++++++++++++++++++++++++++++++++++++++++++++++++++++++++++++++++++++++++++++++++++++++++++++++++++++++++++++++++++++++++++++++++++++++++++
```

Participant: Subject, IdP
Message type: SubjectEnrollmentkey.java (2.4.24)

```
++++++++++++++++++++++++++++++++++++++++++++++++++++++++++++++++++++++++++++++++++++++++++++++++++++++++++++++++++++++++++++++++++++++++++++++++++++++

@javax.annotation.Generated(value = { "ModelPro SoaML
Cartridge" }, date = "2011-10-25T14:52:36.366+01:00")
public interface SubjectEnrollmentkey {
      public String getEnrollmentSubjectRef();
      public void setEnrollmentSubjectRef(String _value);}

++++++++++++++++++++++++++++++++++++++++++++++++++++++++++++++++++++++++++++++++++++++++++++++++++++++++++++++++++++++++++++++++++++++++++++++++
```

Participant: Subject, IdP
Message type: SubjectEnrollmentkeyPart.java (2.4.25)

```
++++++++++++++++++++++++++++++++++++++++++++++++++++++++++++++++++++++++++++++++++++++++++++++++++++++++++++++++++++++++++++++++++++++++++++++++

@javax.annotation.Generated(value = { "ModelPro SoaML
Cartridge" }, date = "2011-10-25T14:52:36.366+01:00")
public class SubjectEnrollmentkeyPart implements Serializable
{
      private String enrollmentSubjectRef;
      public String getEnrollmentSubjectRef() {
            return enrollmentSubjectRef; }
      public void setEnrollmentSubjectRef(String _value) {
            this.enrollmentSubjectRef = _value; }}


++++++++++++++++++++++++++++++++++++++++++++++++++++++++++++++++++++++++++++++++++++++++++++++++++++++++++++++++++++++++++++++++++++++++++++++++
```

Participant: Subject,SP,IdP
Message type: DigitalIdentityRequest.xsd (1.1.2)

```
++++++++++++++++++++++++++++++++++++++++++++++++++++++++++++++++++++++++++++++++++++++++++++++++++++++++++++++++++++++++++++++++++++++++++++++++

<xsd:import
namespace="http://modeldriven.org/examples/message/contractag
reement/schema"
            schemaLocation="ContractAgreement.xsd" />
      <xsd:complexType name="SpIdPDigitalIdentity__Part">
            <xsd:sequence>
                  <xsd:element name="contract"
type="ContractAgreement2:ContractAgreement" minOccurs="1"
maxOccurs="1" />
```

```
                <xsd:element name="sPRef" type="xsd:string"
minOccurs="1"
                      maxOccurs="1" />
            </xsd:sequence>
      </xsd:complexType>
      <xsd:complexType name="SpIdPDigitalIdentity">
            <xsd:sequence>
                  <xsd:element
name="SpIdPDigitalIdentityPart"
type="DigitalIdentityRequest9:SpIdPDigitalIdentity__Part"
minOccurs="1" maxOccurs="1" />
            </xsd:sequence>
      </xsd:complexType>
      <xsd:complexType name="ToConsentResponse__Part">
            <xsd:sequence>
                  <xsd:element name="consentToSend"
type="ContractAgreement2:ConfirmationType" minOccurs="1"
maxOccurs="1" />
                  <xsd:element name="sPRef" type="xsd:string"
minOccurs="1"
                        maxOccurs="1" />
            </xsd:sequence>
      </xsd:complexType>
      <xsd:complexType name="ToConsentResponse">
            <xsd:sequence>
                  <xsd:element name="ToConsentResponsePart"
type="DigitalIdentityRequest9:ToConsentResponse__Part"
minOccurs="1"
                        maxOccurs="1" />
            </xsd:sequence>
      </xsd:complexType>
      <xsd:complexType name="Consent__Part">
            <xsd:sequence>
                  <xsd:element name="contract"
type="${nsPrefixes.get($type2.package)}:ContractAgreement"
minOccurs="1" maxOccurs="1" />
                  <xsd:element name="idPRef"
type="xsd:string" minOccurs="1" maxOccurs="1" />
            </xsd:sequence>
      </xsd:complexType>
      <xsd:complexType name="Consent">
            <xsd:sequence>
                  <xsd:element name="ConsentPart"
type="DigitalIdentityRequest9:Consent__Part" minOccurs="1"
maxOccurs="1" />
            </xsd:sequence>
      </xsd:complexType>
      <xsd:complexType name="DigitalIdentity__Part">
            <xsd:sequence>
                  <xsd:element name="identityattributes"
type="xsd:string"
                        minOccurs="1" maxOccurs="1" />
            </xsd:sequence>
      </xsd:complexType>
      <xsd:complexType name="DigitalIdentity">
            <xsd:sequence>
                  <xsd:element name="DigitalIdentityPart"
type="DigitalIdentityRequest9:DigitalIdentity__Part"
minOccurs="1" maxOccurs="1" />
            </xsd:sequence>
      </xsd:complexType>
```

```
++++++++++++++++++++++++++++++++++++++++++++++++++++++++++++
++++++++++++++++++++++++++++++++++++++++++++++++++++++++++++
+++++++++++++++++++++++++++++++
```

Participant: Subject
JaxWS:ProfileToChallengeReceiverWebService.java(2.1.14)

```
++++++++++++++++++++++++++++++++++++++++++++++++++++++++++++
++++++++++++++++++++++++++++++++++++++++++++++++++++++++++++
+++++++++++++++++++++++++++++++

@javax.annotation.Generated(value = { "ModelPro SoaML
Cartridge" }, date = "2011-10-25T14:52:36.366+01:00")
@Stateless(name = "profileToChallengeReceiver", mappedName =
"Subject/profileToChallengeReceiver", description = "Web
service implementation for port ProfileToChallengeReceiver
defined participant Subject")
@WebService(serviceName = "Subject", endpointInterface =
"org.modeldriven.examples.servicearchitecture._jaxws.ProfileT
oChallengeReceiver")
@HandlerChain(file = "modelpro.jaxws.handlers.xml")
public class ProfileToChallengeReceiverWebService {
      // Participant
      private Subject subject;
      // The POJO implementation of the port type
      private ProfileToChallengeReceiver implementation;
      @PostConstruct
      void init() {     subject = Subject.getInstance();
implementation = subject.getProfileToChallengeReceiver();
implementation.setProfileToChallengeSender(new
org.modeldriven.examples.services.profiletochallenge._jaxws.P
roxy());    }
      /**@param message*/
      public void sendProfile(
org.modeldriven.examples.message.profiletochallenge.schema.Pr
ofile_Schema message) {implementation.sendProfile((message ==
null) ? null
                  :
FlattenedXmlObjectAccessor.newXmlObjectAccessor(
      org.modeldriven.examples.message.profiletochallenge.Pro
file.class, message));}
      // Wrapper operation
      public void sendProfile(
org.modeldriven.examples.services.profiletochallenge.schema.p
rofiletochallengereceiver.SendProfileRequest request)
{his.sendProfile(request.getMessage());
      }
      /**@param message*/
      public void updatedProfileConfirmation(
org.modeldriven.examples.message.contractagreement.schema.Con
firmationType_Schema message) {
implementation.updatedProfileConfirmation((message == null) ?
null: FlattenedXmlObjectAccessor.newXmlObjectAccessor(
org.modeldriven.examples.message.contractagreement.Confirmati
onType.class,message));}
      // Wrapper operation
      public void updatedProfileConfirmation(
org.modeldriven.examples.services.profiletochallenge.schema.p
rofiletochallengereceiver.UpdatedProfileConfirmationRequest
request) {
this.updatedProfileConfirmation(request.getMessage());      }}
```

```
+++++++++++++++++++++++++++++++++++++++++++++++++++++++++++++
+++++++++++++++++++++++++++++++++++++++++++++++++++++++++++++
++++++++++++++++++++++++++++++++
```

Participant: SP
JaxWS:ProfileToChallengeSenderWebService.java(2.1.15)

```
+++++++++++++++++++++++++++++++++++++++++++++++++++++++++++++
+++++++++++++++++++++++++++++++++++++++++++++++++++++++++++++
++++++++++++++++++++++++++++++++

@javax.annotation.Generated(value = { "ModelPro SoaML
Cartridge" }, date = "2011-10-25T14:52:36.366+01:00")
@Stateless(name = "profileToChallengeSender", mappedName =
"ServiceProvider/profileToChallengeSender", description =
"Web service implementation for port ProfileToChallengeSender
defined participant Service provider")
@WebService(serviceName = "ServiceProvider",
endpointInterface =
"org.modeldriven.examples.servicearchitecture._jaxws.ProfileT
oChallengeSender")
@HandlerChain(file = "modelpro.jaxws.handlers.xml")
public class ProfileToChallengeSenderWebService {
      // Participant
      private ServiceProvider serviceProvider;
      // The POJO implementation of the port type
      private ProfileToChallengeSender implementation;
      @PostConstruct
      void init() {serviceProvider =
ServiceProvider.getInstance();
implementation =
serviceProvider.getProfileToChallengeSender();
implementation.setProfileToChallengeReceiver(new
org.modeldriven.examples.services.profiletochallenge._jaxws.P
roxy());}
            /**@param message*/
      public void profileToUpdateRequest(

      org.modeldriven.examples.schema.Profile_Schema message)
{implementation.profileToUpdateRequest((message == null) ?
null
                        :
FlattenedXmlObjectAccessor.newXmlObjectAccessor(
org.modeldriven.examples.Profile.class, message));    }
      // Wrapper operation
      public void profileToUpdateRequest(
org.modeldriven.examples.services.profiletochallenge.schema.p
rofiletochallengesender.ProfileToUpdateRequestRequest
request) {
this.profileToUpdateRequest(request.getMessage());    }
      /**@param message*/
      public void profileRequest(
org.modeldriven.examples.message.profiletochallenge.schema.Pr
ofileProperties_Schema message)
{implementation.profileRequest((message == null) ? null
                        :
FlattenedXmlObjectAccessor.newXmlObjectAccessor(
org.modeldriven.examples.message.profiletochallenge.ProfilePr
operties.class, message));    }
      // Wrapper operation
```

```
        public void profileRequest(
    org.modeldriven.examples.services.profiletochallenge.schema.p
    rofiletochallengesender.ProfileRequestRequest        request)
    {this.profileRequest(request.getMessage());     }}


    ++++++++++++++++++++++++++++++++++++++++++++++++++++++++++++
    ++++++++++++++++++++++++++++++++++++++++++++++++++++++++++++
    ++++++++++++++++++++++++++++++++
```

Participant: Subject Interface:ContractAgreementReceiver.java(2.3.1)

```
    ++++++++++++++++++++++++++++++++++++++++++++++++++++++++++++
    ++++++++++++++++++++++++++++++++++++++++++++++++++++++++++++
    ++++++++++++++++++++++++++++++++

    @javax.annotation.Generated(value = { "ModelPro SoaML
    Cartridge" }, date = "2011-10-25T14:52:36.366+01:00")
    public interface ContractAgreementReceiver {
          /**@param message */
          public void toServeRequest(

          org.modeldriven.examples.message.contractagreement.ToSe
    rve message);
          /**@param message*/
          public void rectificationContractRequest(
    org.modeldriven.examples.message.contractagreement.ContractRe
    ctification message);
          public void contractAgreementRequest();
          /**@param message*/
          public void contractAgreementidPSpec(
    org.modeldriven.examples.message.contractagreement.ContractAg
    reement message);}


    ++++++++++++++++++++++++++++++++++++++++++++++++++++++++++++
    ++++++++++++++++++++++++++++++++++++++++++++++++++++++++++++
    ++++++++++++++++++++++++++++++++
```

Participant: Subject Interface:ContractAgreementSenderver.java(2.3.2)

```
    ++++++++++++++++++++++++++++++++++++++++++++++++++++++++++++
    ++++++++++++++++++++++++++++++++++++++++++++++++++++++++++++
    ++++++++++++++++++++++++++++++++

    @javax.annotation.Generated(value = { "ModelPro SoaML
    Cartridge" }, date = "2011-10-25T14:42:46.505+01:00")
    public class ContractAgreementSender {
          // used interfaces for this port
    private ContractAgreementReceiver contractAgreementReceiver;
      public ContractAgreementReceiver
getContractAgreementReceiver() {
            return contractAgreementReceiver;}
      public void setContractAgreementReceiver(
                  ContractAgreementReceiver _usedInterface) {
            this.contractAgreementReceiver = _usedInterface;}
```

```
    /**<!--begin-user-doc --> <!--end-user-doc -->
     * @modifiable */
    public void sPSubjectDigitalIdentityRequest(){}
    /**<!--begin-user-doc --> <!--end-user-doc -->
     * @modifiable
     * @param message */
    public void sendContract(
    org.modeldriven.examples.message.contractagreement.Cont
ract message)
    {}}
```

Reference

1. ModelDriven.org (2009), *ModelPro*. Available: http://portal.modeldriven.org/project/modelpro. Accessed 28 Oct 2011

Part III
Conclusion and Outlook

Chapter 7
Conclusion and Outlook

Do not spy on one another (49:12) Do not enter any houses except your own homes unless you are sure of their occupants' consent (24:27).

(Holly Quran Verses)

Privacy is becoming an important issue and its importance will continue to grow over time. In the header of CNIL web site,[1] it is mentioned that "information technology must respect the human identity, the human rights, privacy and liberties". We believe that technology is far to be our enemy and can play a key role as a tool to protect human identity, the human rights, privacy and liberties if it is well implemented and used.

7.1 Main Contributions and Summary Conclusions

The digital society has had an important impact on our lives and common society's yardsticks have changed including the concept of identity and privacy. We identified and detailed main issues related to privacy and digital identity and give an overview of some relevant economic and ethical related issues that are faced by individuals, private and public institutions. We analyzed technical issues without forgetting to take into consideration the importance of economic, legal and ethical challenges. We are convinced that non-technical issues are as important as technical ones. We explained that dealing with digital identity and privacy is a complex problem with several facets and for this reason it should be apprehended in a global perspective through a coherent, integrated and multidisciplinary approach. We provided taxonomy of digital identity management definitions based on definition-focus.

[1] http://www.cnil.fr/english/

G. Ben Ayed, *Architecting User-Centric Privacy-as-a-Set-of-Services*, Springer Theses, DOI: 10.1007/978-3-319-08231-8_7,

We studied a comparison between centralized and federated technical models based on set of criteria and we elected federated one and specifically, we explained the supremacy of user-centric identity federation.

Digital identity is increasingly being more persistent, which implies loss of user's control over identity, security risks and threads to privacy. We assumed that digital identity regroups a set of linked and disparate documents distributed over computing ecosystems' domains. Currently, metadata are being democratized and used for various purposes. We suggest an innovative approach based on metadata management, which would weak links between digital identity documents in order to make them less visible, which would foster trusted partnership, and therefore encourage trusted collaboration among networked computing ecosystems. An XRD-based implementation of digital identity document metadata is provided and explained. We extend this work into Content Centric Internetworking environment.

With the emergence of service-oriented economy, distributed systems and cloud computing, many software industry experts and evaluators are encouraging the development and adoption of service orientation and open standards as a mean to assure security and privacy interoperability. In this context, how could we implement interoperable privacy related to digital identity? It is recognized that technical initiatives, emerging standards and protocols are not enough to guarantee resolution for the concerns surrounding a multi-facets and complex issue of identity and privacy. A technical approach is not sufficient enough to tackle privacy issues. A multidisciplinary and integrated approach dictates that law, policies, regulations and technologies are to be crafted together. It is demonstrated that privacy should be incorporated from the very outset of the project. Thus, we began with a specification of business interoperability, through the definition of DigIdeRP requirements that are drawn from global, domestic and business-specific privacy policies. We designed DigIdeRP Framework for organization's security implementation team in order to be able to provide technical interoperability, through the adoption of open standards and implementation of a set of services and service's interfaces: Privacy-as-a-Set-of-Services (PaaSS) that could accommodate any SOA. The framework relays on the idea that privacy requirements should be taken into consideration from the beginning of PaaSS development project. We suggested DigIdeRP framework to help aligning digital identity-related privacy initiatives with the organization's business goals and security strategy. The framework focuses primarily on providing interoperability by disassembling DigIdeRP requirements into services that can integrate an SOA. It clearly define a roadmap that Enterprise/Information System security implementation team, which brings together IT security architects, designers, developers, and analysts, to be able to disassemble DigIdeRP requirements into autonomous, granular and loosely coupled set of services and build PaaSS system. DigIdeRP Framework will serve as a basis for vital understanding between business management and technical managers on digital identity related privacy initiatives. PaaSS enables on-demand privacy; whenever a party is in need of one or multiple elements of DigIdeRP, he could invoke the associated service or services to respond to his need. Thus, PaaSS would inevitably resolve complexities and issues associated with different and various siloed DigIdeRP implementations within identity systems and enable interoperable DigIdeRP to be carried within multiple distributed environments. The layered DigIdeRP framework presents five practical layers as an ordered sequence as a basis of DigIdeRP project roadmap, however, in practice, there is an iterative process to assure that each layer supports effectively and enforces requirements of the adjacent ones. Each layer is composed by a set of specific activities. Specifically, the framework is divided into five layers and three mapping gateways: (1) purpose-level SOA is concerned with articulating the purpose and motivations of the project. The purpose is to build digital identity-related Privacy-as-a-Set-of-Services that could accommodate any SOA. More specifically, the focus of the framework is designing services for SOA and not designing an SOA itself. In the purpose-business mapping gateway, we looked for sources such as privacy policies, procedures, fair information practices and project-specific needs in order to identify DigIdeRP requirements further in the next level; (2) business-level SOA deals with specifying clear DigIdeRP requirements and taking into consideration DigIdM architectural and technical models constraints. In the business-fabric mapping gateway, we identify service candidates' pool from DigIdeRP requirements. Thus, the mapping gateway will facilitate and ease the transition between the two layers; (3) fabric-level SOA copes with identifying and specifying the services, conversation and collaboration between them (interfaces and choreographies), and the way of calling them. In the fabric-platform mapping gateway, we consider several services' deployment environment constraints in the service design such as the component diagrams in UML2 through which we model the transition from business software architecture into technical software architecture; (4) platform-specific-level SOA handles with specific-platform deployment environment of the services such deployment diagram in UML2 that depicts a static view of the run-time configuration; and (5) SOA artifacts-level in which we generate through ready-to-use automatic transformation rules, implementations and codes of SOA artifacts. Completeness of services' implementation that is generated on this level depends on the maturity of the layer4 outputs. We could evolve DigIdeRP Framework to be fully in accordance of model-driven engineering (MDE)/model-driven architecture (MDA) approach. However, we consider purpose-business mapping gateway, business-level SOA, and business-fabric mapping gateway as key elements of MDA Computational-Independent Model (CIM); business-fabric mapping gateway, fabric-level SOA, and fabric-platform mapping gateway as key elements of MDA Platform-Independent Model (PIM); and fabric-platform mapping gateway and platform-specific-level SOA as key elements of MDA Platform-Specific Model (PSM). Inter-and intra-DigIdeRP layers iterations are consequence of SOA delivery lifecycle and strategies alignment. In DigIdeRP Framework, we choose the combination of top-down and bottom-up strategies in a different way from agile approach. The agile strategy allows for the business-level analysis to occur concurrently with service design and development. In the framework, business-level analysis starts to occur in accordance of top–down strategy without going through service design and implementation. Thus, intra-layer iterations occur. However, as far as the top-down progresses, we adopt on-demand bottom-up strategy to allow going back from layers to upper ones. Thus, inter-layers iterations occur. We specified blocks in each layer of

DigIdeRP framework based on OMG SoaML modeling language. The blocks in the framework determine a roadmap that security team could follow to successfully implement PaaSS. SoaML is chosen because it allows technology-neutral representation of services, supports the modeling activities and constructs such as service contracts to properly design collaborative service candidates that could be accommodated by service oriented architecture, fits into an overall model-driven development approach, which is considered as an important aspect because MDA facilitates the design when requirements change, and enables business oriented and systems oriented services architectures to mutually and collaboratively support organization's mission. We detailed blocks in each layer of the framework. In the purpose-level SOA layer, we articulate the need of implementing digital identity-related Privacy-as-a-Set-of-Services. In the purpose-business mapping gateway, we identified the privacy requirements sources related to digital identity such as policies, fair information practices, laws and procedures. We classified them into three groups: (1) privacy business-specific requirements represent the privacy requirements related to identity in particular industry or filed such as banking, health, and education; (2) privacy domestic requirements represent the recurring privacy needs and recommendations related to identity presented by national bodies and local privacy authorities; and (3) privacy global requirements represent the recurring privacy needs and polices related to identity presented by international bodies, regional policy-makers and global legal framework. In addition, it encompasses also the requirements that are neither domestic nor business-specific and the practices and assessment tools that are provided by organizations having a global vision. In business-level SOA layer, we specify four blocks: (1) functional requirements' specification block represents DigIdeRP requirements; (2) DigIdM technical model. DigIdM identity federation is elected because it secures distributed systems and allows to better preserve privacy; (3) DigIdM deployment perspective. DigIdM systems' works and projects have been classified into a landscape of three perspectives: (a) network operator centric perspective in which capabilities that maximize and protect network assets are sought; (b) application service provider centric perspectives in which capabilities that maximize and protect application assets are sought; and (c) user-centric perspective in which capabilities that allow privacy protection and user control over digital identity are sought. Considered as a derivate of DigIdM identity federation, user-centric identity federation is adopted because it provides more control over digital identity; and (4) business process portray. Business Processes Management (BPM) orchestrates the end-to-end business processes and the invocation of services, which in turn may call other services to automate steps in the process. The synergy between BPM and SOA not only allows users better control of the business process as it decoupled from the IT architecture, but also better alignment between Business and IT. We chose to describe DigIdeRP requirements in flow chart-based notation: Business Process Modeling Notation (BPMN 2.0) because process-based description enriches the requirements and after identifying services, process-based description will provide the way how services will be consumed and invoked in order to fully execute the process. Six DigIdeRP processes are identified: (1) ServiceRequest Process; (2) ProfileToChallenge Process; (3) EnrollmentRequest

Process; (4) DigitalIdentityToUpdate Process; (5) PeriodicDigitalIdentityToUpdate Process; and (6) EditDigitalIdentity Process.

Service design approach is an inter-layers block. SoaML modeling capabilities support the service "contract-based" and "interface-based" approaches. We had to choose between the two approaches before undertaking activities in the business-fabric mapping gateway, fabric-platform mapping gateway, layer 3, and layer 4. The service-contract approach requires an already established business and collaboration agreement between parties. In the adopted DigIdM identity federation technical model, CoT sets the agreement between parties of the identity federation, thus, service-contract approach was the best-fit.

We identified, designed and implemented in Java (JEE project) and WSDL/XSD a set of seven services deployable within SOA environments. SoaML service architecture and service contracts diagrams were the starting point that helped to specify services. The services are: (1) ContractAgreement service: the subject plays the role of a Sender of ContractAgreement to the SP in order to establish an agreement about contract's conditions and provisions. The SP plays the role of a Receiver of ContractAgreement; (2) ProfileToChallenge service: the Subject has the ability to challenge his profile that it is in hold by the SP. Thus, the Subject plays the role of a Sender and the SP as a Receiver; (3) DigitalIdentityRequest service involves all participants: the Senders are the Subject and IdP; and the Receiver is SP. The SP asks to receive specifications of IdP(s), which transfer to the SP the subject's identity attributes. The transfer is not possible if the subject has not clearly given his consent. The Subject conveys to the IdP the consent about digital identity attributes dissemination to the SP; (4) PeriodicDigitalIdentityToUpdate service involves the IdP, which is the in-charge of the timing process, as a Sender of the attributes' update request to the Receiver: Subject; (5) DigitalIdentityUpdate service allows to describe the Subject's ability to send new digital identity attributes' values to the IdP, which plays the role of the Receiver; (6) Enrollment service involves the Subject as a Sender of digital identity to be added request to the IdP, which receives the request; and (7) EditDigitalIdentity service implicates the IdP as a sender of digital identity attributes to the Subject, who is already expressed willing to edit digital identity. We provided also the services' invocations roadmap for each process. Figure 7.1 sums up how services, services 'interfaces and methods are responding to the ten DigIdeRP requirements. Consumer interface includes methods that are available for other services to call the service and provider interface includes methods that are available for the service itself to call other services.

Finally, digital identity is primarily a question of education—the education of children as well as educators, human resource managers, chief learning officers, employees and policy makers. We need to maintain in good form the digital representation of ourselves, our digital selves just as we do our physical selves with sports and physical education. DigIdeRP is a question of responsibility that everyone should assume for his or her security and happiness in digital life. If we secure ourselves, others will also be secured.

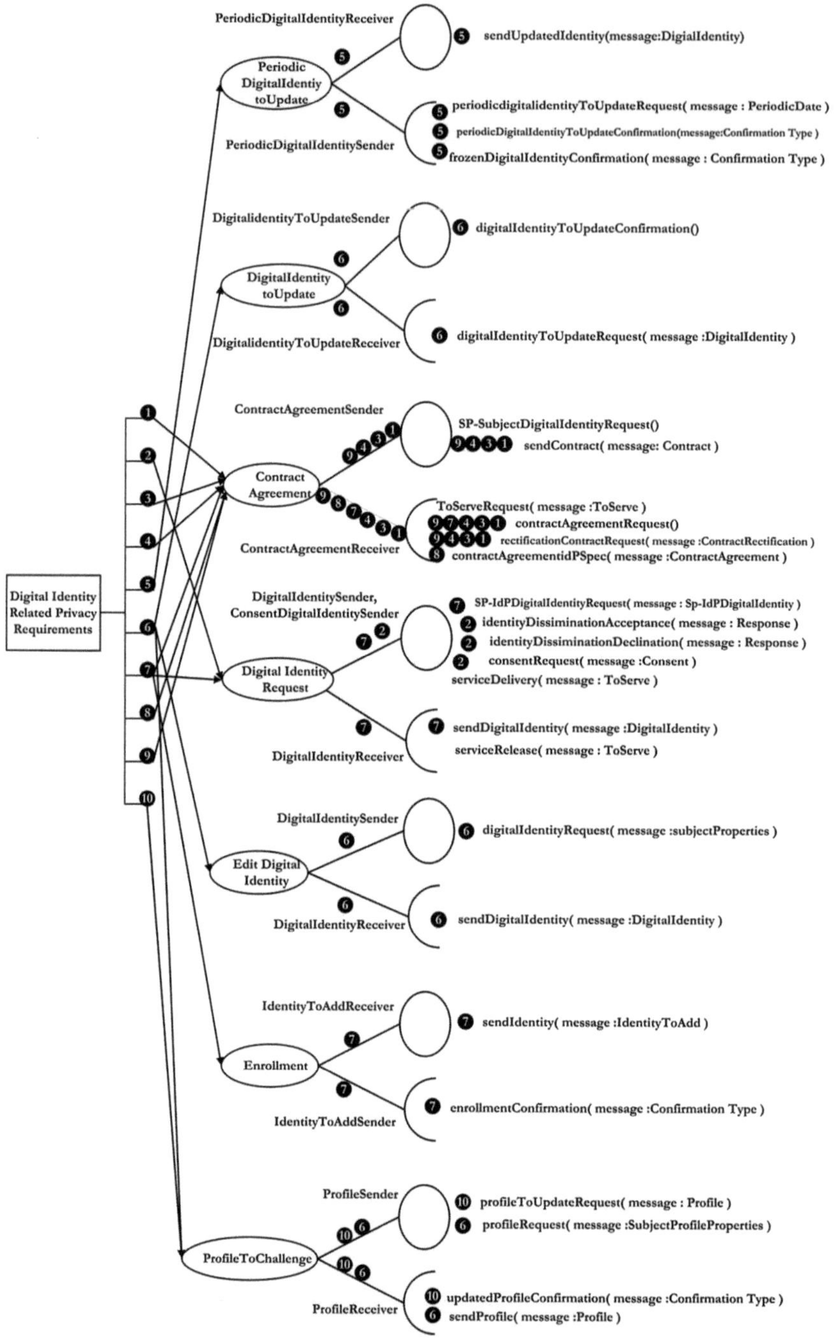

Fig. 7.1 PaaSS system: the implementation of DigIdeRP requirements

When Abraham Lincoln spoke of "a government of the people, by the people, for the people" as a definition of a democracy, he was speaking of a wave of transformation that was changing the way government related to the citizens it served. We think that in our context, we can slightly change the quotation into "a DigIdeRP of the people, by the people, for the people" to stress that DigIdeRP is a responsibility of every person, group of persons, organizations, governments and societies to secure ourselves for the benefit of all stakeholders. User-centricity is also to be taken into consideration when designing DigIdM systems. We believe that this way could be a right wave of transformation.

7.2 Research Limits and Future Work

We cover in this section the limits of the Framework and we explain six directions of work extension and framework improvement that we will follow in the near future.

7.2.1 DigIdeRP Framework Limits and Opportunities of Evolution

DigIdeRP Framework blocks descriptions are based on OMG SoaML, which helps to systemically choose and identify services from services' candidates pool. Services' candidates were elected in ad hoc way. We intend to explore the existence and applicability of other service modeling languages on DigIdeRP Framework and to compare Framework outputs. While SoaML service contracts has provided a major contribution to model DigIdeRP requirements, but we find that it also interesting to explore the development of DigIdeRP with RuleML and to evaluate benefits and inconveniences against possibilities that are offered by SoaML. Rule Markup Language RuleML is a markup language for publishing and sharing rule bases on the World Wide Web. RuleML builds a hierarchy of rule sublanguages upon XML, RDF, XSLT, and OWL [1]; Domain Decomposition Methods [2]; and W3C Unified Service Description Language (USDL), which is a language for describing general and generic parts of technical and business services to allow services to become tradable and consumable [3]. The objective of this research area is to consolidate the framework towards being a modeling-language-independent DigIdeRP Framework. We will implement services from network operator centric perspective and application service provider centric perspective based on the description of each perspective requirements given by ITU [4]. Moreover, we will adopt service interface based approach instead of service contract based approach and we'll explore differences. The major limit of the framework is services longevity issue. When DigIdeRP requirements, DigIdM technical models, deployment or trust models changes, impacts of the changes affect the design and implementation of all services at a risk of existing services reutilization. This is due to the tightly-coupled nature of DigIdeRP requirements. Metamodel for privacy policies within SOA of [5–7] in which researchers have made a decomposition trial of privacy policies, and it is inspiring us to conduct future research to explore whether the hypothesis service identification starts from requirements disassembling rather than from service design is or is not valid. We intend also to generate BEPL code generation either at the intra-service choreography level or at inter-services choreographies (service consumption roadmap, Chap. 5). We need also a terminology consistency when coupling digital identity, privacy and service-orientation. Review of literature and propose taxonomy of definitions including that of as service-oriented identity management, management-led by service, service-led identity management, service-oriented management of identity, identity service management, service-oriented security, service-oriented management of identity, service-oriented approach to identity management, identity-enabled web services, etc. Moreover, we plan to implement services' interfaces security, service calls compliance to WS-security, data security, etc. Finally, we'll look for services description language to catalogue them in a registry.

7.2.2 Service Design and Architecture Metrics

We plan to address design and architectural quality by investigating service design and service oriented architecture quality metrics and processes. These would enrich DigIdeRP framework in order to ensure more quality when designing services. This would help to generate more accurate service implementation. We will explore how a success metric such as Software Maturity Index [8] could help in designing and implementing DigIdeRP services, Value Delivery Modeling Language (VDML) that supports analysis of the development and exchange of values between business parties within a value network or across multiple value networks. The creation of value is often supported by suppliers that provide value in their services [9], and SOA design patterns [10] integration.

7.2.3 PaaSS System Deployment in Service-Oriented Environments

In the European Community report on the future of cloud computing [11], the commission recommends the EC should stimulate cloud computing research and technological development in the area of security, trust and privacy. The emergence of the synergy between SOA and Cloud Computing and the intersection between Cloud Computing and Software as a Service is encouraging us to explore the constraints and the opportunities of deploying PaaSS in the public and private

Cloud Computing environments. Since security, trust and privacy pose issues related to multi-tenancy and control over sensitive data location and non technical issues such as legalistic ones related to intellectual property rights and data protection in the cloud, we find interesting to explore in general the privacy needs in the cloud and specifically DigIdeRP needs in that environment. Is DigIdeRP Framework can help to design PaaSS that could accommodate cloud platforms? Cloud environments involve multiple stakeholders such as cloud providers, cloud resellers or aggregators, cloud adopters software/services vendors, cloud consumers and cloud tool providers. One of the main barriers to implement identity in the cloud is the increased complexity of having to establish trust relationships between enterprises and service providers, while protecting the security and privacy requirements dictated by customers and regulations [12]. In order to be able to deploy at SOA environments such as cloud computing or SOA-compliant PKI, we should carefully look at DigIdM Architectural Models and specify conversations with service-oriented deployment environments such as Platform-as-a-Service (PaaS).

7.2.4 "Forgetting" Persistent Digital Identity and Brain Informatics

Brain Informatics (BI) is an emerging interdisciplinary and multi-disciplinary research field that focuses on studying the mechanisms underlying the human information processing system (HIPS). BI investigates the essential functions of the brain, ranging from perception to thinking, and encompassing such areas as multi-perception, attention, memory, language, computation, heuristic search, reasoning, planning, decision-making, problem-solving, learning, discovery, and creativity. Visionary writings about the Internet often chose metaphors of interconnectivity to describe its potential, many of them borrowed from neuroscience: the "World Brain," a "collective intelligence," and so forth [13]. We plan to extend our work on "forgetting" or making weak links between DigIdDocs by identifying parameters of WeightScore from BI research on brain's forgetting mechanism. In parallel, Nigel Shadbolt and Tim Berners-Lee [14] explain, in their own words, the benefits of studying the Web: "studying the Web will reveal better ways to exploit information, prevent identity theft, revolutionize industry and manage our ever growing online lives". The authors encourage interdisciplinary nature of work such as engineering, biology (plasticity, nervous system), ecology, law, sociology, and medicine fields to better engineer the current and future Web. In the near future, we intend to investigate in details parameters and input variables of GrainScore, DistanceScore, and WeightScore functions. If more variables are identified, a snowflake data schema could be adopted instead of star data schema to reflect the reality of more or less important input variables in DistanceScore function. We'll study also whether this model is applicable into identity federation and user-centricity models.

7.2.5 Digital Identity and Privacy in Content-Centric Internetworking

Evolving from a document-centered into a service and data-centered World Wide Web, Web of data, requires a better user's digital identity protection and management. The promising data centric internetworking capabilities provide a better data recognition and management. The persistent nature of digital identity entails loss of user's control over distributed identity attributes. Digital identity expiration date is one of the identity hiding techniques that we applied to reduce the persistence and give the users' more control over identity attributes. We will take into consideration that expiration date is negotiable between participants of the digital identity federation to establish and reach enough level of agreement upon min and max durations of expiration date with in accordance of permissible expiration date legal, policies, or rules requirements. Called also permissible expiration dates, they represent the contract between disclosers and recipients. And whether is it fixed or variable, permissible expiration dates could reduce "power issue" [15] and gives the user's more control over his digital identity. We intend also to deal with DigIdeRP requirements and their implementation within CCNx open source project [16]. Another consideration would be studying the feasibility of integrating XRI scheme [17] of identifiers instead of CCN content names. XRI provides abstract identifiers that aim to provide a universal format for abstract, structured and platform-independent identifiers, so they can be shared across any number of domains, directories, and interaction protocols. In addition, XRI syntax supports peer-to-peer addressing that allows any two network nodes to assign to each other XRIs and perform cross-resolution.

7.2.6 Digital Identity Management in Data Superabundant Era

Thomas P. Clancy, Vice President of Education Services with EMC Corporation insists in his words: "Not only are we in an information age, we're in an age where information is exploding into a digital universe (…) Just to give you an idea of the challenges we face today, in one year the amount of digital information created, captured, and replicated is millions of times the amount of information in all the books ever written" [18]. Moreover, during 2009, Americans received around 24 years' worth of video footage from aircrafts that flew over Iraq and Afghanistan. The quantity of information in the world is rising. Mankind created 150 EB (2^{60} bytes) of data in 2005 and it is estimated to create 1,200 EB in 2010. According to the Economist article, "storing the bits that might be useful is difficult enough. Analyzing it, to spot patterns and extract useful information, is harder still" [19]. Alex Szalay, an astrophysicist at Johns Hopkins University, notes that the proliferation of data is making them increasingly inaccessible [20]. While Digital memories raises the issue of digital identity persistence but proliferation DigIdDocs would make them increasingly inaccessible, which would foster forgetting capabilities. What are the issues and opportunities provided by digital memories? And how could we secure digital identity and ensure privacy in such environments? We think that they are few critical questions that are worth to respond in the near future.

References

1. RuleML Initiative, The rule markup initiative. (2011), Available: http://ruleml.org/. Accessed 15 June 2011
2. Domain Decomposition Methods. Available: http://www.ddm.org/. Accessed 15 Nov 2011
3. W3C, Unified Service Description Language (USDL). (2010), Available: http://www.w3.org/005/Incubator/usdl/wiki/Main_Page. Accessed 13 Nov 2011
4. ITU Focus Group on Identity Management (FG IdM), Report on identity management use cases and gap analysis (2007)
5. D.S. Allison et al., in *ICITST '09 Privacy and trust policies within SOA* in International Conference for Internet Technology and Secured Transactions (2009)
6. D.S. Allison et al., in *IWSESS '09 Metamodel for privacy policies within SOA*, in The 2009 ICSE Workshop on Software Engineering for Secure Systems (2009)
7. D. Garcia et al., in *ICDIM '10 An Electronic Contract Model for Privacy Protection in Service-Oriented Architecture* in 5th International Conference on Digital Information Management, (Thunder Bay, Canada, 2010)
8. D.S. Herrmann, *Complete Guide to Security and Privacy Metrics (*Auerbach Publications, Florida, 2007)
9. Object Management Group Inc. (OMG), Value Delivery Modeling Language (VDML). (2011), Available: http://neffics.eu/wp-content/uploads/2011/06/11-05-11-VDM.pdf. Accessed 13 Nov 2010
10. T. Erl, *SOA Design Patterns* (Prentice Hall, New Jersey, 2009)
11. L. Schubert et al., The Future of Cloud Computing: Opportunities for European cloud computing beyond 2010 (2009), Available: http://cordis.europa.eu/fp7/ict/ssai/docs/cloud-report-final.pdf. Accessed 21 May 2010
12. M. Dixon, Identity management trends and predictions (2009), Available: http://blogs.oracle.com/identity/entry/identity_management_trends_and_predictions. Accessed 15 Nov 2011
13. R. Galloway and E. Thacker, *The Exploit: A Theory of Networks* (University of Minnesota Press, Minneapolis, 2008)
14. N. Shadbolt, T. Berners-Lee, Web science emerges. Sci. Am. Mag. **299**, 76–81 (2008)
15. V. Mayer-Schönberger, *Delete: The Virtue of Forgetting in the Digital Age* (Princeton University Press, New Jersey, 2009)
16. Palo Alto Research Center, CCNx open source project (2011), Available: http://www.ccnx.org/. Accessed 9 Nov 2010
17. OASIS, An Introduction to XRIs (Working Draft), ed, (2005)
18. T.P. Clancy, in *Forward of the Book, in Information Storage and Management: Storing, Managing, and Protecting Digital Information*, ed. by G. Somasundaram, A. Shrivastava (Wiley Publishing, New Jersey, 2009)
19. K. Cukier, The data deluge: businesses, governments and society are only starting to tap its vast potential. The Economist (23 Feb–5 Mar 2010). Available: http://www.economist.com/opinion/displaystory.cfm?story_id=15579717. Accessed 13 May 2010
20. K. Cukier, Data, data everywhere. The Economist (23 Feb–5 Mar 2010). Available: http://www.economist.com/specialreports/displaystory.cfm?story_id=15557443. Accessed 13 May 2010

About the Author

Dr. Ghazi Ben Ayed is a scientific researcher in the field of cyber-security, he's committed in conducting interdisciplinary research towards making the digital world a safer and a forgiving place. He has many years of progressive teaching, academic and research experiences in North of America, Europe, Africa, and Asia. He holds a Doctorate in Information Systems from University of Lausanne, Switzerland; M.Sc. degree in E-commerce (major: Informatics) from University of Montreal, Canada; Management/Leadership graduate certificates from McGill University, Montreal, Canada and Bachelor in Business Informatics from University of Tunis, Tunisia. He authored multiple publications in ACM, IEEE, and Springer; and he transferred knowledge and diffused cutting edge results to research community and enterprises. In the beginning of his career he played ERP consulting key roles in several projects in Canada and USA and as R&D developer. G. Ben Ayed's current research interests include digital identity, privacy, service-oriented security, digital reputation, trust, Big Data and digital identity hiding.

G. Ben Ayed, *Architecting User-Centric Privacy-as-a-Set-of-Services*, Springer Theses,
DOI: 10.1007/978-3-319-08231-8, © Springer International Publishing Switzerland 2014

MIX
Papier aus verantwortungsvollen Quellen
Paper from responsible sources
FSC® C105338

If you have any concerns about our products,
you can contact us on
ProductSafety@springernature.com

In case Publisher is established outside the EU,
the EU authorized representative is:
Springer Nature Customer Service Center GmbH
Europaplatz 3, 69115 Heidelberg, Germany

Printed by Libri Plureos GmbH
in Hamburg, Germany